About Island Press

Island Press, a nonprofit organization, publishes, markets, and distributes the most advanced thinking on the conservation of our natural resources—books about soil, land, water, forests, wildlife, and hazardous and toxic wastes. These books are practical tools used by public officials, business and industry leaders, natural resource managers, and concerned citizens working to solve both local and global resource problems.

Founded in 1978, Island Press reorganized in 1984 to meet the increasing demand for substantive books on all resource-related issues. Island Press publishes and distributes under its own imprint and offers these services to other nonprofit organizations.

Support for Island Press is provided by Apple Computers, Inc., Mary Reynolds Babcock Foundation, Geraldine R. Dodge Foundation, The Educational Foundation of America, The Charles Engelhard Foundation, The Ford Foundation, Glen Eagles Foundation, The George Gund Foundation, William and Flora Hewlett Foundation, The Joyce Foundation, The J. M. Kaplan Fund, The John D. and Catherine T. MacArthur Foundation, The Andrew W. Mellon Foundation, The Joyce Mertz-Gilmore Foundation, The New-Land Foundation, The Jessie Smith Noyes Foundation, The J. N. Pew, Jr., Charitable Trust, Alida Rockefeller, The Rockefeller Brothers Fund, The Florence and John Schumann Foundation, The Tides Foundation, and individual donors.

RACE
TO SAVE
THE TROPICS

RACE
TO SAVE
THE TROPICS

Ecology and Economics
for a Sustainable Future

Edited by
Robert Goodland

ISLAND PRESS

Washington, D.C. □ *Covelo, California*

Library of Congress Cataloging-in-Publication Data

Race to save the tropics : ecology and economics for a sustainable future / edited by Robert Goodland.
p.   cm.
Includes bibliographical references (p.
ISBN 1-55963-039-6. — ISBN 1-55963-040-X (alk. paper)
1. Environmental policy—Economic aspects—Developing countries.
2. Tropics—Economic conditions.   3. Rain forest ecology—Economic aspects—Developing countries.   I. Goodland, Robert J. A., 1939–   .
HC59.72.E5R33   1990
333.7′0913—dc20   89-71677
CIP

Printed on recycled, acid-free paper

Manufactured in the United States of America
10   9   8   7   6   5   4   3   2   1

IN MEMORIAM

Francisco Mendes Filho
"Chico Mendes"

Brazilian rubber-tapper leader, assassinated
on December 22, 1988. He gave his life
uniting indigenous peoples,
the landless poor, and conservationists
against the destruction of the Amazon rainforest.

CONTENTS

Foreword

SUSTAINABLE SOCIETY THROUGH APPLIED ECOLOGY: THE REINVENTION OF THE VILLAGE

For millennia humans existed as collections of more or less related individuals, living by traditions generated by their surroundings and previous traditions. Life was a relatively sustainable management recipe for the manageable resources, and disaster plans for the unmanageable. Each new human, largely a stack of blank diskettes, was programmed by his or her surviving neighbors, and by some collective memory of those who did not survive. Sustainable agroecosystems were the dominant nonwild vegetation type.

Those social units that did not live in this way suffered the consequences until traditions evolved or they were displaced. The natural world was enemy and friend, storehouse and earthquake. I am not describing a noble savage so much as the observant, scarred, and scared rural human. Berbers, Incas, Dyaks, Yanomamo, Masai, and Toltecs were not created by god(s) but by the equilibrium created by humanity's dominant trait—replacement of wildlands with human domesticates, including humans—checked by the realities of overextension.

And then we added wire and microscopes and airplanes and universities and big boats and doctors and diesel and pesticides and radios and passports and electricity and computers and guns and libraries and chain saws and banks and commercial debt and penicillin and credit cards and bulldozers and nylon and stock exchanges and . . . to the human toolbox. But it's the same individual humans, not changed one whit; it is their collective face and fate that has changed. Short-term sacrifice for the long-term collective good has gone out of style. Humanity has shrugged off its immediate feedback loops, tropical nature has lost the battle, and now *we* are left with a choice. Do we salvage the vanquished? And we are left with a task—get ourselves to where we can perceive the negative feedback before the momentum crushes the objective.

The single largest specific task facing tropical humanity today is reinventing the traditions that adjust population size and demands to the size and nature of the resource base. What makes this reinvention so excruciatingly difficult is that, today, the consequence of trashing the natural resource base is either ameliorated or surfaces much later (and thus is not experienced by the trasher). Humanity is living on a credit card. Every time the going gets rough, we just change address. Those that have the ability to invent new traditions can leave the scene and those that cannot, die. Not the best of circumstances for evolution.

What are the basic ingredients of evolution? Selective processes that favor one tradition over another, a variety of traditions among which the selective process chooses, and a mechanism by which the favored traditions are transmitted. We have had nearly two decades of global public exposition of the burgeoning and every day more indiscriminate selective processes. As this book emphasizes, it is time to move on past advertising the presence of selection. Let's move on to increasing the variety of traditions among which the selective process can select, and get to thinking about the transmission and institutionalization of this information into traditions once again. Far more than just local agroecological traditions must appear and replicate; the quickest way to get many pesticides out of tropical rivers may be to cause the acceptance of blemished bananas on the New York market.

Each of the component parts of the natural resource base—water, soil, biodiversity, wood, air, heat, rain, and so on—has gone through and is going through the above process, often at different rates and perceptions. Focus on biodiversity for a moment. For millennia, it is just there—the huge gene pool from which the occasional Indian grandmother selects a new crop plant, the rope-maker selects a new fiber, the shaman selects a new heart-stopper. It is omnipresent, exuberant, hardly used by a local culture, mostly unknown. It almost always reseeds itself when trashed (except for the megafauna swept off the New World continents by early humans, the 100-plus species of birds eaten off the Hawaiian Islands by the early canoeing colonists, the hundreds of species of plants barbecued off the Australian dry topics by 50,000 years of burning, etc.). Agricultural inviability, diseases, and the certainty that all foreigners are thieves and killers maintained the balance.

Then humanity escapes from its village, and the tropical world is converted to a global village unshackled by feedback for half a millennium. The global neighborhood is converted to Lebanon, El Salvador, and the Oklahoma dust bowl. The most conspicuous casualty is the multitude of traditions, and the selection for them, that once put crops in the ground in the right season, left trees standing on village edges, put 46 varieties of rice in one backbreaking field, and kept the local population at a size that matched the resource base. Lying in the open grave with tradition is biodiversity. As crushed as were European wheat fields, playgrounds for Allied and Axis tanks. We suffer, and live with, the consequences of our actions.

Traditions will rise again. The process that generates them—memory perceived by the next generation—is not dead. But biodiversity must be lifted from the grave, planted back out, nursed a bit, and explicitly left in peace. Humanity is not about to give the earth back for another 4 billion years of evolution, and biodiversity has lost the greatest friend it ever had—agricultural and social inviability for most of the tropics. Perhaps one of the most delicate of all traditions—the incorporation of biodiversity's intellectual stimulation in the village social and mental fabric—is about to be taken from the three-quarters of humanity that live in, and always will live in, the tropics. A million square kilometers of intensive agriculture can be highly sustainable agroecology, and the most deadly boring green cocoon in the universe. The mental health of the tractor driver, mayor, and mother is as much an ingredient in a sustainable agroecosystem as are integrated pest management, polyculture, and reforestation. The destiny of humanity is not glorious fields full of human draft animals picking beetles off bushes. Biodiversity conserved in wildlands has far more function than its very great value as a global gene bank and phytochemical cornucopia.

This book calls for the application of ecology; the irony is that ecology ever got unapplied. This book highlights, pleads for, attention to the technology of sustainability in natural resource use. The irony is that human sustainability has always been a combination of planning and restrictive reality, molded by feedback. Humanity simultaneously escaped from both the reality and the feedback. All the return to sustainability really requires is voluntary—more fun than involuntary—submission to traditions of sustained natural resource management before global selection does it for us at a much lesser steady state value. No noun is regal; context is. A monoculture rice field may be sustainable resource management, or it may be a disaster. Corporation agriculture may be the best management structure on the north side of the mountain, family farms the best on the south side, and no farm at all to the east and west. A tropical rain forest may be best restored 200 meters from a metropolitan area and a remote rain forest best turned to macadamia nut plantations. Cows belong in some national parks.

Which is all to say that humans also have to decide what really is the resource that they want. Does anyone really want to live in an oil palm plantation, a sugarcane field? When humans have the option, it is pretty clear that they take the advantages of modern society and decorate the margins with biodiversity and unsullied water, air, views, and food. Someone challenged me on that once—he said that the oil palm plantation worker would rather have a TV set than a national park nearby. Who said either/or?

I don't argue for a return to the tropical village. But what is crystal clear is that the site-specific traditions that were once held together by god, marriage, and pests must now be replaced by a new species of social glue that once again renders the human occupant responsible for its actions, be they underfoot or at the end of a 5,000-kilometer fax line. Our diskettes are receiving an amazing assortment of miscellaneous programs, but we have slipped out from beneath the editor's thumb and we are sending our printouts to random addresses. You are bright? Then put your mind to the generation of programs that will base the fitness of the individual once again on the fitness of the village and its habitat. These programs will be applied ecology, by definition.

DANIEL H. JANZEN

PREFACE

"The most important challenge for tropical ecology is for tropical ecologists to stop intellectualizing and start implementing what needs to be done." Thus wrote Dan Janzen (1988), arguably the world's foremost practitioner of applied tropical ecology. This book shows what applied tropical ecology is and how it already has started to be used in improving economic development in developing countries.

Applied ecology, the application of ecological principles to solving real world problems, is discussed in the introduction. The integration of applied ecology into the process of economic development has already improved it significantly and will improve it much more when such integration becomes systematic rather than the exception it is at present. Even so, applied ecology is an intractable science as Slobodkin (1988) points out: "One reason is that ecology is called upon to solve all types of real and conceivable problems about the interaction of organisms and their environment. . . . Both ecology and medicine when confronted with questions about new symptoms attempt to establish ad hoc research programs to find solutions, and thereby tend to annoy clients who want quick, direct replies."

This book, written by front-line, muddy-footed practitioners of applied ecology, shows the enormous benefits to be gained and the staggering and usually irreversible losses to be prevented when ecology is applied to economic development. Ecology is applied to agricultural development, both agroecology and agroforestry, moist forest management, pest management, and to large-scale hydro and irrigation schemes, all of them in tropical developing countries. This is complemented by chapters on teaching applied ecology to nationals of tropical countries; ecology applied to conservation; and environmental sustainability.

Applied tropical ecology is a young science—this is the first book to be published on the subject. This book is not a research tome. Rather, it records the attempts, many empirical and ad hoc, to apply what little is already known to real life development issues in tropical countries.

RACE
TO SAVE
THE TROPICS

INTRODUCTION

ROBERT GOODLAND AND MARYLA WEBB

Economic development has been vigorously pursued as the means for achieving the goal of fulfilling human potential and improving standards of living. Economic development is a complex process, usually accompanied by major cultural changes involving systems of governing, social relationships, attitudes toward nature, and technological artifacts. Development is usually characterized by a gradual transformation from a subsistence economy—in which families or villages are essentially self-sufficient, and any trade of goods and services is largely through barter—to a market economy—in which production of goods and services are specialized and trade in goods and services involves a national currency—or by growth in a market economy. Market economies are frequently measured as growth in gross national product (GNP). GNP is a national account statistic that measures aggregate national generation of income or the monetary value of resource flows. Only recorded economic transactions involving a national currency are included in the statistic. Its environmental deficiencies are recognized but not amplified here.

After World War II, attention began to focus on economic development for more-traditional societies in largely tropical countries not yet so affected by the world market economy and modern technological conveniences. This led to international cooperation for economic development on an unprecedented scale. International lending agencies and programs of bilateral aid were established to transfer resources and

3

provide technical assistance to developing countries. The governments of developing countries also set about fostering the market economy—setting macroeconomic policies, investing to build infrastructure and expand basic services, fostering and sometimes engaging directly in productive activities. Although these efforts have been successful in transforming millions of lives and increasing per capita GNP—per capita GNP has risen an average of 3 percent annually since 1965 in developing countries—it can be argued that overall quality of life and human well-being has not risen commensurately.

This chapter suggests how and why economic development has not always been successful in increasing quality of life. It focuses the impact of economic development on the natural environment—although the impact of economic development on the social environment also is a critical issue and lack of attention to it has caused social malaise and failure in specific development projects. Also, it is suggested that applied ecology can enhance economic development so that it preserves rather than damages quality of life and of the environment; the influence of multilateral development banks (the World Bank, founded in 1945; the Inter-American Development Bank, founded in 1959; the African Development Bank, founded in 1964; and the Asian Development Bank, founded in 1966) on the economic development process is discussed. How applied ecology can enhance these multilateral development banks is described, and critical areas for future research are suggested.

ECONOMIC DEVELOPMENT AND QUALITY OF LIFE

The acute environmental stresses many developing countries experience are intensifying rapidly. Overall many of today's activities are degrading the world environment at an accelerating rate.

More environmental degradation and destruction has probably occurred in the last twenty years—while GNP was rising 3 percent a year in developing countries—than the previous fifty. The greatest environmental destruction in the history of humanity has occurred in the last one hundred years, as technologies have become more powerful, agriculture has become more mechanized, and the market economy has spread globally. Despite laborious attempts to solve environmental problems, implemented solutions to deforestation, desertification, overfishing, pollution, the greenhouse effect, acid rain, depletion of the ozone layer, and extinction of species evade us. Daly and Cobb (1989) offer the most encouraging ways out of current planetary impasse.

One of the reasons for this state of affairs is that the very process of economic development leads to environmental degradation; whereas traditional systems of resource use (social and technical) often set up circumstances whereby environmental quality is preserved. For example, traditional systems usually develop through long periods of trial and error in an area. They build detailed familiarity and sensitivity to a local environment. The ones that preserve environmental quality are often essentially subsistence systems (including storing modest surplus for lean times). The people have little incentive to overexploit the environment, since most traditional societies—even migratory ones— are tied to a certain territory and therefore have a stake in its preservation and protection so that it continues to provide for societal needs. In many traditional societies, preserving nature or respecting the land is a universal human obligation.

Conversely, in modern developed societies, most people lose sensitivity to, understanding of, and appreciation for their natural environments. In the world market economy, the incentive is to produce as much surplus as one can sell, because stockpiling wealth increases materials power and options. Nothing in the structure of the market induces conservation. On the contrary, if the demand is there, all incentives are to overexploit in modern societies. The profits from sustainable harvesting are less per unit time than unsustainable "mining" of a potentially renewable resource. If one resource becomes depleted, all that is necessary is to have stockpiled enough capital to allow substitution of another resource. Fiscal policies, such as tax (dis)incentives, can soften this fundamental market characteristic, but cannot remove it. Treatment of the land that provides for all one's material needs (except for one's immediate living space) becomes no longer a particular concern. Modern societies usually take a usability approach to nature—that is, they do not adhere to a belief in a moral obligation to nature.

ROLE OF APPLIED ECOLOGY

The systematic application of ecology (page 6) to development planning, although not a panacea, can go a long way toward alleviating the problem outlined. Although much remains to be discovered, ecology

has uncovered many principles of natural ecosystem functioning critical
to successful development planning. Most failed economic development projects have ignored critical ecological attributes of an area. At
the very least, they are not likely to achieve as many sustainable benefits
as economically as would be achieved if ecological principles were used.

Applied Ecology Defined

Applied ecology is the application of ecological principles to solving real world
problems. Theoretical (pure) ecology uncovers the ecological principles that
explain the structure and functioning of the natural world. Both branches of
ecology are essentially positive, rather than normative, activities. They focus on
what is or will be, rather than on what ought to be.

Even so, applied ecology is closely associated with normative activities. A
broad political goal, or the solution to a local problem—how things ought to
be—is usually decided in the political or socioeconomic arena. Applied ecologists use basic ecological knowledge to aid reaching socioeconomic or political
goals. These goals include the conservation of biological diversity or sustainable natural resource management such as the optimization of products from
forests. Applied ecologists also conduct research at a local level to solve a
particular problem, such as the alleviation of an algal bloom in a lake, or the
establishment of the boundaries of a nature reserve.

Whereas theoretical (pure) ecologists are concerned with the natural world
for the sake of knowledge, applied ecologists are concerned with applying that
knowledge in order to achieve human goals or solve human problems. Pure and
applied boundaries overlap. An ecologist who normally works on theoretical or
empirical problems may address a more immediate, pressing problem. Conversely, an ecologist working on a more applied problem may uncover information that advances the science. Moreover, much theoretical or empirical work
helps solve real world problems, or has as its general aim the eventual better-
ment of humanity.

SOURCE: Walker and Norton, in press.

The systematic use of ecological principles in development planning
is beneficial (it increases the quality of life) in four main ways: by
fostering the productivity of the natural resources upon which all development depends; by favoring the maintenance of environmental quality; by promoting efficient and sustainable natural resource use; and

by avoiding unexpected negative consequences of actions. These are arbitrary categories. A healthy and productive ecosystem will have high environmental quality. Efficient and sustainable natural resource use tends to preserve both productive ecosystems and environmental quality.

The use of ecological principles in development planning fosters environmental health and productivity. This is true in any type of development project that manipulates or affects the natural world in order to achieve human goals. This principle is illustrated in the impoundment of rivers (page 8). Through ecological research and planning before a dam is constructed, the health and productivity of these ecosystems can be preserved while other energy or agricultural goals are simultaneously achieved.

The use of ecological principles in economic development also favors the maintenance of environmental quality in forest resource use. For example, ecologists have developed ecologically based, forest-management systems that preserve the structure and functioning of tropical forests. These are usually long-term, sustained-yield silvicultural schemes based on encouraging the natural regeneration of commercially valuable tree species. Many of these systems are taken from the knowledge of native foresters and indigenous or tribal peoples, and include selective logging, pre-felling vine-cutting, directional felling, and care in extraction—largely by using balloon-tire vehicles, blimps, and carefully located and engineered roads.

Lack of efficient and sustainable economic development often stems from failure to address ecological information in project design. Billions of dollars can be lost. Two, among many, vivid examples concern failure to appreciate the intricacies of pollination ecology. The first example concerns the Brazil nut (*Bertholletia excelsa*) industry. Before ecological investigations, we did not understand that this commercially important tree depends upon Euglossine bees as a major pollinator, as well as agoutis—large forest-dwelling rodents—for seed dispersal. Euglossine bees, in turn, depend upon pheromones from specific epiphytic orchids for mating signals; and the orchids, in turn, depend upon certain forest conditions for growth. This helps explain why Brazil-nut plantations have commonly failed and investors have lost millions of dollars; one or more links in this interdependent web were omitted. Use of this ecological knowledge in planning would have ensured that all parts of the system were present, preventing major losses of scarce capital. Similarly, lack of ecological knowledge led to the loss of billions of dollars on foregone yields and the enormously expensive six decades of hand pollination when the oil palm was introduced from West Africa to Malaysia in 1917 without knowing that it is insect pollinated. The West African pollinator, the weevil *Elaeidobius kamerunicus*, was introduced into Malaysia 64 years later, in 1981. This boosted the next year's yield alone by 130,000 tons or 57 million U.S. dollars in foreign exchange.

Ecological Effects of Dams

The main effects of dams on downstream riverine ecosystems commonly include (1) reduction in species diversity of the whole spectrum of plants, insects, and fish including microflora and fauna; (2) explosive population increases of pest species including disease-carrying insects and river-choking aquatic weeds; (3) degradation of the river channel including decreased stream-bank stability; (4) increased fish mortality, particularly of juveniles; and (5) decreased recruitment of juveniles into the adult breeding population resulting in declining fish numbers. The main floodplain effects can include: (1) an increase or decrease in salinization of floodplain pans; (2) invasion of the riparian zone by nonriparian plants that have little tolerance to flooding (this increases the devastating impact of a severe flood); (3) alteration of floodplain reptile, bird, and mammal species as some species are eliminated or decrease in population size. Estuaries, especially in arid regions, are sensitive to impoundment flooding to scour the mouth and open it to the sea again. Effects on estuaries can include the following: (1) permanent closure of the estuary; (2) build-up of sand- and/or mud-flats; (3) decreased interactions between the estuary and its associated lagoons and salt marshes; (4) changes in species composition of estuary and associated land areas, including reductions in the number of angling fish and bait organisms and dramatic increases in nuisance organisms such as mosquitoes.

Source: Ward et al., 1984.

The improvement of agricultural practices through applied ecology is probably the most critical task facing humankind today. As the cost of agricultural input has soared, problems with biocide poisoning and groundwater contamination have become epidemic, loss of topsoil has intensified, and destruction of natural ecosystems and wildlands continues, the efficiency and sustainability of current capital-intensive farming methods have been brought into serious question. Agroecologists and others have been designing alternative agricultural schemes that require fewer inputs and are less destructive of the natural world. Many agroecologists are investigating the agricultural practices of traditional peoples and are finding that traditional farming methods often make ecological sense. Others have created research farms that test innovative practices based on knowledge of the natural world (see below). Rodale (1981) found that yields using regenerative agricultural practices, although initially lower, burgeoned as the land recovered from previous abuses. Although yields have not yet quite reached the levels of the more capital-intensive methods, overall profits are greater because of decreased expenditures on inputs. More output per input equals increased efficiency. Since input usually depends on limited fossil energy, this translates into increased sustainability.

Finally, the use of ecological principles reduces unexpected environmental costs. An example of unnecessary damage to environmental quality that occurred through avoidable lack of attention to ecological principles is a 1949 midge-control program. Clear Lake, California, is a beautiful resort, well-known for fishing and water-skiing, but vacationers were annoyed by midges. Midges are closely related to mosquitoes. Though they are neither bloodsucking nor dangerous to human health, midges were bothersome because of their large numbers. Ten years of spraying DDT resulted in: a DDT-resistant strain of midge, which is still a nuisance; elimination of the western grebe from the lake; reduction in the population of other water birds; and poisoning of the fish, the lake's most important asset. It is ironic that the midge problem was caused by human activities in the first place. Nutrients draining into the lake in the runoff from farmlands and in raw sewage dumped directly into the lake had changed the bottom sediments to favor the growth of the midge larvae. The midges were also attracted to the increasing numbers of lighted houses near the edge of the lake. By using this information about the ecology of the system, the midge could have been controlled without the ecological backlashes.

Another example of long-lasting damage to environmental productivity brought about by avoidable lack of attention to ecological principles is Amazonian cattle pasture development. During the last decade, large tracts of Amazonian rain forest have been felled and burned to establish cattle pasture. Most of these enterprises have failed because most nutrients are depleted after the first few years of cattle production, and weeds impair the pasture. Maintaining production proved impossi-

One Example—Biological Pest Control

An example of wise use of ecological knowledge in development problem-solving is the biological control program of the cassava mealybug (*Phenacoccus manihoti*) and green spider mite (*Mononychellus* spp.) being conducted by the International Institute of Tropical Agriculture (IITA) based at Ibadan, Nigeria. This mealybug–spider-mite complex has severely decreased cassava yield in recent years and has spread to 30 African countries, causing an estimated annual $2 billion in damages. The goal was to control these pests and to restore potential production levels of this formerly important source of food. The ecological knowledge that cassava and its pests were balanced in Latin America, where these species evolved with the introduced pests, failed because pests had become too numerous or the timing prevented effective control agents. Test releases of carefully chosen natural enemies of both the mealybug and the mite have been very successful. One predator of the mealybug, the parasitoid *Epidinocarsis lopezi*, maintained itself in the experimental field throughout the following rainy season and kept the mealybug under control in the dry seasons of 1982–1984, while spreading over 5 kilometers into neighboring farms. The same predator, released at a second test site, successfully colonized a 3,000-hectare mealybug-infested cassava farm, and beyond, has been one of the most efficient for achieving the objectives for several reasons: once achieved, it has been self-sustaining, i.e., it is maintained naturally without any recurrent costs to the farmer; and it is specific, i.e., it appears to achieve only the objective, without any side effects such as damage to the health of the farmer, nontarget wildlife species, or the ecosystem. Moreover, it will prove to be extremely cost effective—most costs will be up front in the research and testing. There are no apparent health or environmental costs. And subsistence farmers will not have to continuously buy large supplies of biocides that would have been required as the only alternative. Hence, the pest will not build up resistance to the biocide, as is usually the case, and there will be no need to continuously develop and use new biocides.

SOURCE: IITA, 1984.

ble without fertilizers. Abandoned pastures are still essentially barren and likely to remain so. Tropical forests recover after moderate distur-bance, such as slash-and-burn agriculture, if certain physical and biolog-ical elements are present. Bulldozing or high-intensity cattle ranching destroys those elements, however, and precludes the natural recuperative power of the forest.

The Fukuoka Farm

An example of high farm output for minimal input, as promoted by Rodale and other low-input farmers, was worked out by Mr. Masanobou Fukuoka, a microbiologist turned farmer in Japan. Mr. Fukuoka has perfected a system of farming that requires no plowing, little weeding, and no agricultural chemicals, yet produced yields comparable to modern capital-intensive methods. With essentially no ecological training, but astute powers of observation and immense patience, he used trial and error to eliminate, one by one, any unnecessary agricultural practices. Once he managed to tilt conditions in his fields and orchard slightly in favor of his crops, he interfered as little as possible with the plant and animal communities. As he explains, "My way of farming now is little more than sowing seed and spreading straw, but it has taken me 40 years to reach this simplicity." Although Mr. Fukuoka may not be consciously applying textbook ecological principles, his method makes sense to any agroecologist. Moreover, although this is only one temperate country example of the efficiency that can be achieved using "ecological" principles, it may represent one of the best.

SOURCE: Fukuoka, 1978, 1985.

MULTILATERAL DEVELOPMENT BANKS IN INTERNATIONAL ECONOMIC DEVELOPMENT

Multilateral development banks (MDBs) have a critical role to play in fostering the use of applied ecology so that the process of economic development achieves the goal of improving quality of life, i.e., maintaining environmental quality. One reason the role of these institutions is so critical is their pervasive influence on economic development in developing countries due to the enormous amounts of capital that these banks control. For example, in 1984 the four MDBs together lent $20 billion U.S. currency to developing countries; this was nearly 30 percent of all international development assistance to the public sector in that year. The World Bank represents the largest share of this lending; it lent more than $17 billion U.S. currency in fiscal year 1987 alone. As a comparison, the total GNP of most developing countries is less than $10 billion U.S. a year and total government expenditures range from $100 million U.S. to $3.5 to 4 billion at most.

This ratio of MDB assistance to all international development assistance, large as it is, understates the power of MDBs. The influence of dollar figures on development policies and priorities is magnified by: (1) the much larger counterpart sums provided by recipient countries (commonly 70 percent of total project costs), other development agencies, and private banks; (2) the major flow of technical assistance; (3) involvement in a country's institution-building; (4) the leverage of loan conditionality both in specific projects and in structural adjustment loans; (5) the integral research role in country analysis and planning that influences lending priorities; and (6) the leverage in conjunction with the International Monetary Fund in determining a borrower's creditworthiness to private banks.

MDBs need applied ecologists in two general areas: formulation of overall national development strategies and policies; and the planning of environmentally positive development projects. Applied ecology is beginning to be used in project planning but is not yet systematic in formulation of national development strategies. In 1987, the World Bank created seven new environmental divisions, four in the four geographic regions (Africa, Asia, Latin America, and the rest) and three in a policy planning and research department. Most of these divisions may have an applied ecologist.

Applied Ecology in National Planning

The benefits of using ecological principles in country-wide planning are beginning to be recognized. Besides the more general reasons discussed earlier for using applied ecology in the economic development process, whether on a national or project level, there are also unique reasons for applying ecology to national planning. These reasons all relate to the utility of addressing environmental concerns comprehensively, rather than piecemeal. The essentially intersectoral and interdisciplinary nature of environmental concerns makes this approach fundamental. The benefits of such a strategic approach include the following: improved integration of environmental quality and economic concerns at a national level, i.e., ensuring that the planning actions and policies work in tandem to achieve the goals of economic progress and preservation of environmental quality; improved intersectoral cooperation, since many environmental problems can be traced to poor coordination between sectors, e.g., agriculture and fisheries, forestry and livestock; improved recognition of environmental constraints to, or opportunities for, various forms of development. All of these will result in improved planning and quality of specific projects. Five national-level approaches that address environmental concerns in a more systematic and comprehensive way are now outlined. These approaches can only be effective to the extent to which they are integrated into national planning and backed with budgets and policy changes. Too often, attempts to use some of these approaches have not been so integrated and therefore have fallen short of their goals.

NATIONAL CONSERVATION STRATEGIES. Most attempts by governments to apply ecological principles in national planning have used a blueprint called the *World Conservation Strategy,* prepared in 1980 by the International Union for the Conservation of Nature and Natural Resources in cooperation with the United Nations Environment Programme and the World Wildlife Fund. It was the collaborative effort of more than 700 scientists and advisers associated with these organizations; 450 other government agencies and conservation organizations commented on the document as well; it was formally endorsed by the World Bank. These guidelines aid national planners to integrate ecological principles into the development process for the reasons just discussed. Government leaders, usually in conjunction with aid agencies or nongovernmental organizations, tailor this blueprint to prepare national or regional conservation strategies suited to their unique environmental, social, political, and economic situation. While approaches and methods vary considerably, the principal intent in each case is to improve economic development planning through the timely inclusion of relevant environmental concerns into the planning process. The World Conservation Strategy has five main objectives: the maintenance and proper functioning of the earth's essential ecological processes—ranging from the global hydrological cycle to the local insect or bat pollination of flowering plants, including crops—and life support systems—roughly equivalent to major ecosystem types, e.g., forests, wetlands, grasslands; the preservation of the world's genetic diversity, i.e., conservation and protection of the earth's many species of plants and animals; the protection of renewable resources by controlling levels and patterns of use within sustainable limits; the maintenance of environmental quality; and the achievement of a steady state of resource use and population numbers at a reasonable standard of living and quality of life.

In order to achieve these objectives, *National Conservation Strategy* assesses the current environmental situation, including obstacles to achieving the goals, and sets forth policies, major programs, and specific projects designed to overcome the obstacles and achieve the objectives. Implementation schedules and accountable institutions are drawn up for all recommended actions. Monitoring systems are established so that plans can be adapted to changing circumstances. Since March 1987, 23 countries had prepared or were preparing national conservation strategies (Table 1).

ENVIRONMENTAL PROFILES. The development of an environmental profile is a similar exercise to the development of a national conservation strategy. Both are planning documents that point out the main features and problems of a given country's environment, and recommend specific actions. Environmental profiles, however, have no formal relationship to the world conservation strategy and may not include all its objectives, although their objectives overlap. The purpose of an environmental profile is "to provide a comprehensive picture of a country's environmental and natural resources, including an analysis of the present use, management and conservation, and natural resources, such as soil, water, trees, forests, and rangelands . . ." (IIED, 1986). Environmental profiles also serve to promote recognition of the environment as a system of interacting components and to highlight crucial gaps in environmental knowledge.

A phase I profile is a desk study that summarizes available data on soils, vegetation, water resources, wildlife, and protected areas. A phase II profile is a comprehensive in-country study that may conclude with recommended strategies or actions needed to deal with any environmental problems uncovered. Environmental profiles have been funded by the United States Agency for International Development, sponsored by UNESCO's Man and the Biosphere Program. Many were prepared by the International Institute for Environment and Development (IIED). The Asian Development Bank has also prepared environmental and natural resources profiles for some of its developing member countries as part of its country programming process. In both agencies, these profiles are used in planning country development strategies as well as specific projects. As of March 1987, 51 developing countries had

TABLE 1

**DEVELOPING COUNTRIES WITH NATIONAL
CONSERVATION STRATEGIES**

Region/Country	Status
Sub-Saharan Africa	
Botswana	IP
Guinea-Bissau	IP, Proposal F (1985)
Ivory Coast	IP, Proposal F (1983)
Madagascar	E, F
Senegal	IP, Proposal F (1984)
Sierra Leone	IP, Proposal E (1985)
Togo	IP, Proposal F (1985)
Uganda	IP
Zaire	IP
Zambia	E (1984, 1985)
Zimbabwe	IP (1988)
Asia/Near East/North Africa	
Bangladesh	IP
Indonesia	IP
Jordan	(1)
Malaysia	(2)
Nepal	IP, Prospectus E (1983)
Pakistan	IP
Philippines	IP
Sri Lanka	IP (1988)
Tunisia	IP
Vietnam	E, F (1985)
Latin America/Caribbean	
Barbados	IP
Belize	IP

IP, in preparation; E, available in English; F, French.
(1) Looking for funding.
(2) Some studies for particular states are available.

SOURCE: IIED (1986) and Lists of USAID Country Environmental Profiles and National Conservation Strategies, Sectoral Library, the World Bank.

completed phase I profiles, and 16 phase II profiles (Table 2). Starting in 1987, the World Bank has now completed environmental issues papers for many of its borrowing member countries.

TABLE 2

DEVELOPING COUNTRIES WITH PHASE I OR II ENVIRONMENTAL PROFILES

Region/Country	Phase I Profile	Phase II Profile
Sub-Saharan Africa		
Burkina Faso	(1980)	E (1982)
Burundi	(1981)	
Cameroon	(1981)	
Cape Verde Is.	(1980)	
Gambia	(1980)	
Ghana	(1980)	
Guinea	(1983)	
Lesotho	(1982)	
Liberia	(1980)	
Malawi	(1982)	
Mali	(1980)	
Mauritania	(1979)	
Niger	(1980)	
Rwanda	(1981)	
Senegal	(1980)	
Sudan	(1982)	
Swaziland	(1980)	
Uganda	(1982)	
Zaire	(1980)	E,F (1981)
Zambia	(1982)	
Zimbabwe	(1982)	
Asia/Near East/North Africa		
Bangladesh	(1980)	
Burma	(1982)	
Egypt	(1980)	
India	(1980)	
Jordan	(1979)	
Morocco	(1981)	
Nepal	(1979)	
Oman	(1981)	

Region/Country	Phase I Profile	Phase II Profile
Pakistan	(1981)	
Philippines	(1980)	
Sri Lanka	(1978)	
Syria	(1981)	
Thailand	(1979)	
Tunisia	(1981)	
West Bank/Gaza	(1980)	
Yemen	(1982)	
Latin America/Caribbean/ Central America Region		IP
Barbados	(1982)	
Belize	(1982)	E
Bolivia	(1979)	E (1980) (S-IP)
Costa Rica	(1981)	E,S (1982)
Dominican Republic		E,S (1981)
Ecuador	(1979)	S (1983)
El Salvador	(1982)	S (1985)
Guatemala	(1979, 1981)	S (1948, Exec. Sum. E)
Guyana	(1982)	
Haiti	(1979)	IP
Honduras	(1981)	E,S (1982)
Jamaica	(1981)	IP
Nicaragua	(1981)	
Panama	(1980)	S (1980)
Paraguay		E,S (1985)
Peru	(1979)	S (1986)

SOURCE: As in Table 1.

NATURAL RESOURCE INVENTORIES. A natural resource inventory is a detailed list of the extent and condition of a country's or region's renewable natural resources, and may include forests, fisheries, wetlands, and grasslands. Most inventories are compiled for sectoral planning, such as for the forestry sector. Some natural resource information is monitored annually for inclusion in a government's statistical yearbook. Cropland or soil resource inventories are conducted for this purpose at least. Many natural resource inventories for a particular country or region are a byproduct of other studies. A global tropical forest inventory done by the Food and Agriculture Organisation (FAO) and UNEP using LANDSAT imagery has given many developing countries a wealth of information on forests, as well as other natural resources. Few countries or regions in the developing world have comprehensive, multi-sectoral natural resource inventories.

Even so, a detailed natural resource inventory is a powerful planning tool. In addition, knowledge of the status and trends of resource stocks can allow time to prevent problems before they reach crisis proportions. For example, knowledge of a likely timber short-fall can allow time for preventive steps, such as augmented plantations. Moreover, natural resource inventories, like environmental profiles, provide a bench mark as a basis for evaluating trends in natural resource developments so that remedial measures can be taken as soon as possible.

ENVIRONMENTAL SECTOR REVIEWS. A number of countries are using environmental sector reviews as an important input to their national development plans. Indonesia undertook an extensive environmental review as part of its formulation of the fourth five-year plan (Repelita IV). This detailed planning exercise was supported by the United Nations Development Programme (UNDP), UNESCO, the International Institute for Environment and Development (IIED), and the governments of Belgium, the United Kingdom, and the United States. In a similar vein, Malaysia's third five-year economic-development plan addressed environmental concerns nationwide for the first time (with assistance from the World Bank). This included establishment of an environmental ministry.

YEAR 2000 OR 21ST-CENTURY STUDIES. The Global 2000 study was conducted over a three-year period largely by the U.S. Council on Environmental Quality, the Department of State, and other appropriate U.S. agencies. It described the probable changes in the world's population, natural resources, and environment through the end of the century. It was the first U.S. government effort to look at the interrelations of all three factors from a long-term global perspective. Recognizing the utility of this study, Canada, China, and Korea soon requested assistance from the U.S. government to pursue similar long-term studies of the interaction of population, natural resources, and environmental trends within their borders. An organization, the Global Studies Center, was formed in Washington, D.C., to assist these nations and others in planning and executing their own studies. It was headed by Gerald Barney who headed the Global 2000 study.

More than thirty "Year 2000," "21st Century Studies," or similar studies were under way or have been completed by various governments or other organizations around the world (Table 3). Although these studies vary in the depth and extent of analysis, all present valuable resource-trend information and insights for national planners on strategies for affecting the future of peoples and nations.

ENVIRONMENTAL STATISTICS AND THE CALCULATION OF GNP. Many economists argue that currently used national-income accounting systems, particularly GNP (see page 3), are inadequate for development planning because environmental factors are undervalued or ignored. In particular, GNP can be increased by depleting renewable natural resources, by debasing environmental quality, or by attempts to compensate or correct for these two activities. The kinds of economic activities that increase GNP, but are actually attempts to compensate for decreased environmental quality, include medical expenditures on diseases related to environmental abuse (e.g., bronchitis from air pollution), expenditures to clean up polluted or degraded environments, expenditures to enhance degraded environments (e.g., expenditures to save an endangered species or reforest a watershed), and defensive expenditures to protect ourselves from pollution or other hazards of modern living. These kinds of expenditures actually reflect decreases in national welfare, since the same capital could have been spent on truly productive activities. Therefore, national planners who use GNP to gauge success in promoting the well-being of their people may be inadvertently misled into promoting policies and activities that are actually damaging the nation. For this reason, four UNEP/World Bank Environmental Accounting Workshops (1983, 1984, 1985, 1986) linked environmental accounting and economic accounting. This method increases the quality of feedback to development planners so that activities promoted bring overall benefit to the nation (Ahmad and El Serafy, 1989).

CARRYING CAPACITY. Carrying capacity is a useful applied ecology concept that can be defined as the maximum population of a given species that can be supported indefinitely, allowing for seasonal and random changes, without any degradation of the natural resource base that would diminish this maximum population in the future. It is a

TABLE 3

COUNTRIES AND REGIONS WITH 21ST CENTURY STUDIES

Region or Country	Completed	In Progress
Africa (by Ethiopia and the Ivory Coast	(1983)	x
Caribbean (by Jamaica)		x
Europe (by Austria)	(1977)	
Australia		x
Brazil		
Canada	(1985)	
China	Phase I (1986)	Phase II
Finland		x
Iceland	(1979)	
Indonesia		x
Ireland	(1983)	
Japan	(1983)	
Korea		x
Mauritius		x
Mexico		x
Netherlands	(1978)	
Peru		x
Philippines	x	
Poland		x
S. Africa		x
Sweden		x
Taiwan		x
Tanzania	(1983)	
United Kingdom	(1983)	
U.S.A.	(1982)	
U.S.S.R.		x
Venezuela		
World (by U.S.A.)		x

SOURCE: Global Studies Center, 1611 N. Kent St., Suite 600, Arlington, VA 22209. The Center is a nonprofit charitable and educational organization that assists groups conducting long-term studies of the development and security of individual nations.

familiar concept to wilderness and range managers concerned with protecting the environmental quality of a particular land area. If one of these managers finds that their area of responsibility is becoming degraded, they respond either by culling animals or by increasing carrying capacity (usually temporarily), such as by chemical fertilizer or by planting fodder plants. In the simpler system of a national park or a rangeland, it is more easily recognized that environmental modifications that increase carrying capacity require either importation of materials (e.g., saplings, chemical fertilizers) or energy or the expenditure of energy (e.g., plowing, mowing) to keep the biological system in a state in which it would not naturally stay. In other words, increases in carrying capacity are temporary unless energy (other than solar) is continuously imported; no technological advances can escape this fact.

The concept of carrying capacity can be a useful tool for national planners in protecting the environmental quality of a nation or region. If environmental degradation in a nation is increasing, it may be concluded that the carrying capacity of that nation is being exceeded. National planners can respond by implementing population policies, reducing consumption, increasing efficiency, or importing the materials and energy to increase their carrying capacity temporarily. There is no escaping the fact that, except for the continuous infusion of solar energy, the earth is a closed system. Importation of materials and energy require that they be exported from somewhere that in its turn, creates environmental impacts of its own.

Applied Ecology in Specific Development Projects

Global and national approaches to conservation outlined in the last section will greatly facilitate environmental planning at the project level. If the national development strategy conflicts with conservation goals, then specific projects will also conflict. But no amount of national environmental planning can completely compensate for an environmentally flawed project. Too often, however, this is the role that applied ecologists must play—attempting to reverse the worst predicted impacts of a flawed project. On the other hand, little environmental mitigation is necessary for projects designed so as to maintain environmental values consistent with a national conservation strategy. Therefore, applied ecologists should be involved in the preparation of projects, from the earliest design phase onward.

The role of applied ecology on a project level should ideally follow this sequence: designing and carrying out preproject studies that aid in project design or in predicting environment impact; designing mitigatory/compensatory measures to prevent or compensate for possi-

Dumoga Irrigation Project, Indonesia

An example of a World Bank project that benefited greatly from the involvement of applied ecologists is the Dumoga Irrigation Project in Indonesia. This project promotes irrigated rice on 197 square kilometers of arable land in Dumoga, North Sulawesi, as well as infrastructure, training, and extension. The credits (loans) signed in 1980 provide 1.1 million U.S. dollars (1.5 percent of a total project cost of 71.1 million U.S. dollars) to elevate the status of the 935-square-kilometer nature reserve (established in 1979) into the 3200-square-kilometer Dumoga National Park. This is designed to protect the watershed on which the irrigation depends, to augment dry seasonal flows, to moderate wet seasonal flows, and to reduce the risk of sedimentation into the irrigation infrastructure. Meanwhile, the National Park conserves many endemic forest biota, including the rare giant civet (*Macrogalidia musschenbroeki*) and the endangered Maleo fowl (*Macrocephalon maleo*). Sixty park management personnel have been hired, trained, and equipped. A management plan has been drawn up and is being implemented. The approximately 400 families who arrived after 1981 to live within the park's boundaries have now been relocated. Without the project the National Park may not have been created and would not have received 1.1 million U.S. dollars in start-up costs. Prevailing deforestation of the irrigation catchment would probably have intensified, impairing both the irrigation water supply and the environmental values and services of the forested watersheds.

As a spinoff of this modest investment, this reserve has become host to a seminal scientific expedition. Project Wallace* of the Royal Entomological Society of London (RESL) has been described as "the jewel in the crown of the society's centenary celebrations." Six years in the planning, the year-long expedition to the monsoon forests of Sulawesi, Indonesia, was the world's largest entomological expedition ever mounted. More than 150 scientists from 19 countries took part. The myriad studies addressed a wide range of topics and greatly advanced the knowledge of this unique ecosystem, an area cut off from its neighbors in the Indonesian archipelago for millions of years. Early indications are that the importance of this expedition for the advancement of the biological sciences, including ecology, taxonomy, and medicine, will be similarly large. The project is also expected to have commercial benefits and should produce knowledge that will improve the management of national parks and watersheds in similar tropical moist forest ecosystems.

SOURCES: Knight and Schofield, 1985; RESL and IIS (n.d.).

---

* Alfred Russel Wallace (1823–1913) was president of the Royal Entomological Society from 1820 to 1871, and formulated the theory of evolution with Charles Darwin in 1858.

ble negative effects; implementing environmental measures; monitoring environmental quality to provide early warnings; and evaluating the environmental aspects of completed projects.

PRE-PROJECT STUDIES. Potential ecological impacts of certain classes of projects—water impoundments, irrigation, forestry, livestock, agriculture, highways, industries—are already well known (see below). A knowledge of these potential results is important in improving project design. Knowledge of the particular ecosystem is also necessary. Through preproject studies, an applied ecologist can reduce any gaps in ecological knowledge that may be necessary for successful project design.

DESIGN. With the ecological knowledge gained through both basic research and preproject studies ecologists can create or improve project design to achieve economic as well as environmental values. An environmentally well-integrated project would have these aspects: emulation and accommodation of naturally evolved patterns and processes; emulation or enhancement of successful indigenous or traditional systems of resource use; compatibility with traditional cultural or societal values; usage of species adapted to specific ecosystems in the communities where they evolved; recycling of wastes; usage of solar-based power where possible (should not require major imports of energy); material and energy efficiency; and ecosystem stabilizing (e.g., overharvesting can cause permanent shifts in ecosystem structure and function).

MITIGATORY/COMPENSATORY MEASURES. When, despite best efforts, projects cannot be designed to eliminate most, if not all, negative environmental consequences, then compensatory measures may alleviate them. For example, if a dam inundates some portion of an endangered ecosystem, salvage of all biologically important species, as well as the protection of an equivalent area (both ecologically and spatially) in a park or preserve may be an acceptable compensatory measure.

IMPLEMENTATION AND MONITORING. Implementation of environmental measures is best accomplished by applied ecologists. For example, the establishment of a reserve or park to protect a watershed, or in compensation for the elimination of another wilderness area, requires a thorough understanding of the ecosystems for their setting, design, and management. Likewise, the establishment of biological or integrated pest management requires an understanding of the biology of the pest and prey species, as well as how they interact with each other and the plant host.

The success of an environmental measure, as any other aspect of the project, requires monitoring in order to ensure smooth implementation or to solve unexpected problems. For example, in monitoring a water diversion scheme, it can be discovered that the rate and volume of water withdrawal is destroying an economically and biologically important wetland downstream, despite best efforts to study and design the scheme to prevent that effect. In that case, plans may be redrawn to reduce irrigated acreage in the water-short season in order to preserve the fishery dependent on the intact wetland. Most of the time, trouble-shooting is necessary, as many queries arise in implementing any new enterprise. Applied ecologists keep the environmental aspect of a project running smoothly.

EVALUATION. One neglected aspect of applied ecology is the environmental evaluation of past development projects. Most evaluations of completed development projects (those in which all disbursements have been made on all financed activities, such as dam construction) focus on adherence to disbursement and other schedules and the economic rate of return over the life of the project, which continues after project completion. Success in these two areas is often the yardstick by which project success is measured, even if a few years of project economic success destroyed any future use of the resource base, as well as pertinent environmental values. Environmental aspects of projects are perfunctorily evaluated and the project judged a failure on these grounds only if their neglect has also caused proximal economic failure of the project.

If development is to be truly sustainable, as well as maintaining quality of life, development institutions must devote more resources to the evaluation of environmental aspects of past development efforts. An evaluation should contain a detailed balance sheet of costs and benefits, both economic and environmental, over the long term.

Often, projects fail environmentally because of lack of attention to ecological principles already known. However, occasionally project failure is due to environmental circumstances that could not have been foreseen with then current knowledge. In these instances, only by the careful study of why things turned out differently from what was intended, and the dissemination and use of the results, can past mistakes be avoided in future projects.

COST/BENEFIT ANALYSIS. Current cost/benefit analysis methods or their improper use give inadequate weight to ecological variables and environmental quality issues in project work. Giving the greatest weight to bottom-line financial figures—giving significant advantage to those factors that can readily be quantified and assigned a market value—is a recognized tendency. This is unfortunate for environmental quality because of one main reason easily understood by ecologists: quantification or economic valuation of most ecological values is fraught with practically insurmountable difficulties and in case of ethics may be undesirable. How does one quantify and monetarily evaluate the complex architecture of a rain-forest canopy, the sociobiology of a king vulture, or the mutualisms between Brazil-nut trees, orchids, and agoutis? Attempts to rectify this situation by quantifying and evaluating various environmental benefits and costs are an incomplete solution at best, and no less value laden than a more qualitative approach. (This is because value judgments are required in the choice of studied variables, from a practically infinite pool, and their eventual evaluation.) Moreover, where the course predicted by ecologic versus economic criteria conflicts—as is usually the case in today's world—it can be forcefully argued that it is the economic analysis that needs adjusting. However, one is then in the ridiculous position of having to manipulate economic data to predict the same course already judged prudent through ecologic analysis.

In an ideal world, all projects and activities should be formulated using standard and rigorous ecological and ethical criteria (among others), then evaluated economically only relative to one another. If this procedure is followed by everyone, humanity stands to gain the greatest long-term benefit in terms of quality of life. Violators of these criteria, however, can accrue financial gains during their lifetime at the expense of others and future generations. In this sense, our situation is much like the prisoner's dilemma of sociological/ecological theory. The requirement that projects adhere to certain environmental criteria is an alternative approach that has been suggested before (Baum and Tolbert, 1985; Goodland and Ledec, 1987; Ledec and Goodland, 1989). These criteria can be referred to as safe minimum standards, a nonmarket criteria that a project must meet to be approved. If a proposed project cannot meet these criteria, a more environmentally prudent project should receive scarce investment funds instead. Ecologists are indispensable in formulating and applying site-specific safe minimum standards.

CRITICAL AREAS FOR FURTHER RESEARCH

Since the above activities will be only as good as our ecological understanding, that understanding could be improved by asking the following questions: What is environmental quality? Why should it be maintained? How can it be maintained in the context of economic development?

Some of the confusion and debate about the environmental implications of development activities lie in differing understanding of environmental quality. Most people, particularly in developed societies list as criteria for a high-quality environment: clean air; pure water; healthy, intact soils; and presence of some plant and animal life. However, these criteria could just as well describe an oil palm plantation as a tropical forest. What is missing from these criteria is an appreciation of ecological and evolutionary processes as they determine future life on earth. Janzen (1986) has brought home the difference between managed and natural habitats and the ecological and evolutionary implications of current human management of the environment, so it will not be covered again here.

Fairly pristine natural areas (i.e., wildlands) still possess what Aldo Leopold in his classic, *A Sand County Almanac,* termed ecosystem integrity. Currently, we have only abstract, qualitative descriptions of ecosystem integrity and environmental quality and little understanding of how much one depends on the other. Ecological science should objectively quantify environmental quality. It may sound improbable, but there was once a time not too long ago when we could describe pure water only in qualitative terms. The development of sophisticated chemical techniques now allows us quantitatively to measure the purity of water. Only our ability to measure water purity allows us to legislate water quality. In the same way, until we are able to objectively measure environmental quality, it is difficult to require development to maintain it.

Why should we maintain environmental quality? Again, since there is much misunderstanding about the nature of environmental quality, it follows that there would be similar misunderstanding about why it should be maintained. For example, if environmental quality were defined to be synonymous with ecosystem integrity, there would be (and is) tremendous disagreement about how much of the earth's natural ecosystem can be disrupted and destroyed, and, of those places that we disrupt, how much integrity must be preserved. We have not begun to really answer these questions. In a sense, we are performing the experiment right now. Although in some areas it appears that our intense management is sustainable and causing no major deleterious effects, in other areas the ensuing degradation should tell us we have gone too far. Complete judgment (because of the slow nature of many ecological processes) will not be known for years.

Another interesting, but poorly understood, aspect of why we should maintain environmental quality is the influence of the environment on human physiological and psychological well-being. Much anecdotal evidence suggests that major deviations from the environmental conditions under which human beings and other life forms evolved often impair well-being (see below).

Environmental Effects on Human Well-Being

Examples include the health effects of biocides or heavy metals, where society has created unnatural substances or placed natural substances in the environment in concentrations new to human bodies. Subtler examples are the effects of scenery, sounds, and lighting on the human physiology and psyche. Hospital patients whose windows face trees recover faster than patients whose windows face buildings; and proper exposure to full-spectrum (natural) lighting is linked to healthy physiological and mental functioning (i.e., lack of it can result in or exacerbate learning disabilities or hyperactivity in children, and depression, lethargy, and immune system malfunction in adults). These examples show that the human system is more sensitive to a wider range of environmental factors than previously appreciated, and may require largely natural conditions (conditions that prevailed during the majority of our evolution) for optimal functioning.

Our actions show that we assume we could safely and sustainably convert most of the earth to "corn fields, cow pasture, and cityscapes," figuratively speaking. Today's global society appears to wager that the worst that can happen will be that the earth will be aesthetically poorer (no small thing in and of itself), but otherwise we will just have more money in our pockets and more meat on the table. This may not be a safe bet, however, and increasing evidence substantiates this view. This evidence must be collected and put into a coherent ecological framework. Until it is clear to everyone why the world should not be unraveled so quickly and extensively, we can expect the experiment to continue.

Finally, much more ecological work must be done on how we can maintain environmental quality in the context of economic development. Ecologists cannot just tell development planners what we must maintain and why; we must also tell them how. We cannot just tell them they are doing it wrong, unless we also tell them how to do it right.

In spite of the importance of the role of ecology, Paul Ehrlich (1986) points out that hundreds of well-trained and competent ecologists in the United States alone are restricted in their research by lack of financial support. Jobs in universities, international agencies, governments, and industry do not exist for the thousands of students who wish to become ecologists. Many biological field stations around the world, the bases for research and training in ecology, are essentially nonfunctional for lack of funds. Finally, funds for the preservation of biological diversity—especially through protection of intact ecosystems such as tropical forests—are only a minuscule percentage of what is needed to tackle the task at hand. This alone should justify increased investment in ecological research, training, and institution building as part of economic development.

Ehrlich goes further to state that the reason for this critical state of affairs is the general public's lack of appreciation for the importance of ecology. Humanity has not yet learned how to live within the constraints set by nature, and time is short. "Nature's machinery is being demolished at an accelerating rate, before humanity has even determined exactly how it works. Much of the damage is irreversible. To protect and (to the extent possible) repair the machinery will require many more scientists delving into its complexities, as well as a much deeper appreciation by the general public and decision makers of both its importance and its basic design."

The integration of ecological principles into the economic development planning process is necessary to ensure that such development will achieve the desired goal of sustainably enhancing the quality of life, as well as supplying goods and encouraging production. This integration should occur regionally or nationally, as well as on a project by project basis. Without it, unforeseen ecological backlashes plague efforts to manage and control the natural world. To achieve economic goals while maintaining environmental quality, as pressure for increased production mounts, requires increasing application of ecological principles to development problems. Development can be vastly improved to the extent development planners involve applied ecologists in this process.

References Cited and Sources of Further Information

Abramov, I., 1985. Health effects of interior lighting: discussion. Ann. NY Acad. Sci. 453:365–370.

Ahmad, Y.J., and El Serafy, S., 1989. Environmental accounting. Washington, DC: World Bank, 100 pp.

Barrett, G.W., and R. Rosenberg (eds.), 1981. Stress effects on natural ecosystems. New York: Wiley, 305 pp.

Baum, W.C., and S.M. Tolbert, 1985. Investing in development: lessons of World Bank experience. New York: Oxford University Press, 610 pp.

Berger, T.R., 1985. Village journey. The report of the Alaska Native Review Commission. New York: Hill and Wang, 202 pp.

Carpenter, R.A. (ed.), 1983. Natural systems for development: what planners need to know. New York: Macmillan, 485 pp.

Corry, S., 1984. Cycles of dispossession: Amazon Indians and government in Peru. Survival Internat. Rev. 43:45–70.

Council on Environmental Quality (CEQ) and the U.S. Department of State (USDOS), 1980. Global 2000 report to the President. Washington, DC: Government Printing Office, 3 vols.

Daly, H.E., 1980. Introduction to the steady-state economy. *In:* H.E. Daly (ed.), Economics, ecology, ethics: essays toward a steady-state economy. San Francisco: Miller Freeman Publ., pp. 1–31.

Daly, H.E., 1985. The circular flow of exchange value and the linear throughput of matter-energy: a case of misplaced concreteness. Rev. Soc. Econ. 43(3):279–297.

Daly, H.E., and J.B. Cobb, 1989. For the common good: redirecting the economy toward community, the environment, and a sustainable future. Boston: Beacon Press, 467 pp.

Dasgupta, P., 1982. The control of resources. Oxford: Blackwell, 233 pp.

Davis, S.H., 1988. Indigenous peoples, environmental protection, and sustainable development. Washington, DC: World Bank, 78 pp.

deGraaf, N.R., 1986. A silvicultural system for natural regeneration of tropical rain forests in Suriname. Wageningen, Netherlands: Agricultural University, 250 pp.

di Castri, F., F.W.G. Baker, and M. Hadley, 1984. Ecology in practice. Paris: UNESCO, 2 vols., 920 pp.

Djajadiningrat, 1983. Report of the Environment Sector Review. Jakarta, Indonesia: the Republic of Indonesia and UNDP, Project INS/83/009.

Ehrlich, P.R., 1986. Ecological understanding is not free. Amicus J. (Winter):7–8.

Fukuoka, M., 1978. The one-straw revolution: an introduction to natural farming. Emmaus, PA: Rodale Press, 181 pp.

Fukuoka, M., 1985. The natural way of farming: the theory and practice of green philosophy. Tokyo: Japan Publications, 280 pp.

Goodland, R., and G. Ledec, 1987. Neoclassical economics and principles of sustainable development. Ecol. Modelling 38:19–46.

Goodland, R., C. Watson, and G. Ledec, 1984. Environmental management in tropical agriculture. Boulder, CO: Westview Press, 237 pp.

Goldman, C.R., 1976. Ecological aspects of water impoundment in the tropics. Rev. Biol. Trop. 24 suppl. (1):87–112.

Grainger, A., 1985. Desertification: how people make deserts, how people can stop, and why they don't. London: IIED, 94 pp.

Hardin, G., 1986. The tragedy of the commons. Science 162:1243–1248.

Hecht, S.B., 1982. Cattle ranching development in the eastern Amazon, Forestry dissertation, Berkeley: University of California, 454 pp.

Hinkle, L.E., and W.C. Loring (eds.), 1977. The effect of the man-made environment on health and behavior. Atlanta, GA: Centers for Disease Control, 315 pp.

International Institute for Environment and Development (IIED), 1986. Country environmental profiles, natural resource assessments and other reports on the state of the environment. Washington, DC: IIED.

International Institute for Tropical Agriculture (IITA), 1984. Biological control of cassava mealybug and green spider mite. Ibadan, Nigeria: IITA, 24 pp.

International Union for Conservation of Nature and Natural Resources (IUCN), U.N. Environment Programme (UNEP) and World Wildlife Federation (WWF), 1980. World conservation strategy: living resource conservation for sustainable development. Gland, Switzerland: IUCN, 54 pp.

Jackson, W., W. Berry, and B. Colman (eds.), 1984. Meeting the expectations of the land: essays in sustainable agriculture and stewardship. Berkeley: North Point Press, 250 pp.

Janzen, D.H., 1986. The future of tropical ecology. Ann. Rev. Ecol. Syst. 17:305–324.

Jorgensen, S.E. (ed.), 1983. Application of ecological modeling in environmental management. New York: Sinauer Assoc., 489 pp.

Knight, B., and C. Schofield, 1985. Sulawesi: an island expedition. New Scientist (3 Jan.): 12–15.

Ledec, G., and R. Goodland, 1989. Wildlands: their protection and management in economic development. Washington, DC: World Bank, 278 pp.

Lee, J.A., 1985. The environment, public health and human ecology: considerations for economic development. Baltimore, MD: Johns Hopkins University Press for the World Bank, 288 pp.

Leopold, A. 1966. A sand county almanac. New York: Ballantine Books, 272 pp.

Liepert, C., 1986. Social costs of economic growth as a growth stimulus: the need for a new indicator of net production. J. Econ. Issues 20(1): 109–131.

May, R.M., 1981. Theoretical ecology: principles and applications. Sunderland, MA: Sinauer Assoc., 489 pp.

McKenzie, G., 1983. Measuring economic welfare: new methods. New York: Cambridge University Press, 187 pp.

McGurk, H. (ed.), 1977. Ecological factors in human development. Amsterdam: N. Holland Publ., 295 pp.

Mollison, B., and D. Holmgren, 1978. Permaculture One; a perennial agricultural system for human settlements. Melbourne, Australia: Transworld Publ., 128 pp.

Mollison, B., 1979. Permaculture Two: practical design for town and country in permanent agriculture. Stanley, Australia: TAGARI, 150 pp.

Myers, N., 1980. Conversion of moist tropical forests. Washington, DC: NAS, 205 pp.

National Academy of Sciences (NAS), 1973. Environmental quality and social behavior. Washington, DC: NAS, NRC, 86 pp.

National Research Council (NRC), 1982. Ecological aspects of development in the humid tropics. Washington, DC: National Academy Press, 297 pp.

National Research Council (NRC), 1986. Ecological knowledge and environmental problem solving: concepts and case studies. Washington, DC: National Academy Press, 388 pp.

Norse, E.A., K.L. Rosenbaum, D.S. Wilcove, B.A. Wilcox, W.H. Romme, D.W. Johnston, and M.L. Stout, 1986. Conserving biological diversity in our national forests. Washington, DC: Wilderness Society, 116 pp.

Parrish, J.A., C.F. Rosen, and R.W. Gange, 1985. Therapeutic uses of light. Ann. N.Y. Aca. Sci. 453:356–364.

Pearce, D.W. (ed.), 1975. The economics of natural resource depletion. London: Macmillan, 220 pp.

Pearce, D.W., 1983. The limits of cost-benefit analysis as a guide to environmental policy: a reply to Professor Cooper. Kyklos 36(1):112–114.

Ray, A., 1984. Cost-benefit analysis: issues and methodologies. Baltimore, MD: Johns Hopkins University Press, 158 pp.

Rich, B.M., 1985. The multilateral development banks, environmental policy, and the U.S. Ecol. Law Q. 12(4):681–745.

Ricklefs, R.E., 1982. Ecology (second edition). Concord, MA: Chiron Press, 966 pp.

Rodale, R., 1981. Our next frontier: a personal guide for tomorrow's lifestyle. Emmaus, PA: Rodale Press, 242 pp.

Royal Entomological Society of London (RESL) and Indonesian Institute of Sciences (IIS), n.d. Project Wallace: commemorative expedition to North Sulawesi. January–December 1985. London: Dramrite Printers, 15 pp.

Seymour, J., and H. Girardet, 1986. Far from paradise: the story of man's impact on the environment. London: British Broadcasting Corp., 216 pp.

Smil, V., 1984. The bad earth: environmental mismanagement in China. Armonk, NY: Sharpe, 245 pp.

Timberlake, L., 1985. Africa in crisis: the causes, the cures of environmental bankruptcy. Washington, DC: IIED, 232 pp.

Tolba, M., 1983. Development without destruction: evolving environmental perceptions. Dublin: Tycooly, 197 pp.

Ulrich, R.S., 1984. View through a window may influence recovery from surgery. Science 224(4647):420–421.

Walker, B.H. and G.A. Norton, in press. Applied ecology: towards a positive approach. Johannesburg, South Africa: Centre for Resource Ecology, 46 pp.

Ward, J.V., B.R. Davies, et al., 1984. Stream regulation *in:* Allanson, B.R. and R.C. Hart, (eds.), Limnological criteria for management of water quality in the southern hemisphere, South African National Scientific Programmes Report No. 93. Pretoria: Foundation for Research Development, 181 pp.

World Bank, 1982. Tribal peoples and economic development: human ecologic considerations. Washington, DC: World Bank, 111 pp.

World Bank, 1986. World development report 1986. New York: Oxford University Press, 256 pp.

Worldwatch Institute, 1987. State of the world: a Worldwatch Institute report on progress toward a sustainable society. New York: Norton, 263 pp.

Chapter 1

APPLIED ECOLOGY AND AGROECOLOGY: THEIR ROLE IN THE DESIGN OF AGRICULTURAL PROJECTS FOR THE HUMID TROPICS

STEPHEN R. GLIESSMAN

The long-term sustainability of our world's agriculture has become the topic of considerable discussion and concern among a broad spectrum of people involved in the multiple components of the production, processing, and distribution of food, feed, and fiber around the globe. This is especially true for those who work in agriculture in the humid tropics. Faced with the need to meet the demands both of growing populations and of increasing foreign debt, most tropical countries look to agriculture to feed their people and solve serious cash-flow problems. Based on the experience of development projects of recent decades, though, most agricultural projects in tropical regions have not delivered the promised results.

33

An economic model of analysis for determining the success or failure of agricultural development projects in the tropics has determined to a large extent the approaches that have been taken. Production targets are proposed or set, and various agricultural technologies are applied in attempts to meet those levels. Inputs are increased, based on the knowledge that modern technologies can certainly increase yields. During the past decade we have seen dramatic increases in the costs on the input side of the model, but the value received on the output side does not seem to be rising accordingly. At the same time, we have become aware of serious problems of environmental degradation of air, water, soil, and biotic resources as a direct consequence of many modern farming practices that have been introduced into the tropics from more temperate regions. Reduced returns or even failures of such projects are seen as one of the causes for the increased pressure on tropical rain-forest reserves as massive deforestation continues. But arguments for change in our approach to evaluating agricultural projects based purely on environmental issues has not been enough. We must develop focus on managing agricultural resources in the tropics that both reduces the dependence on costly, nonrenewable inputs and ameliorates the impacts that current practices are having on the environment. By applying an ecological element to the design and management of agricultural projects, as well as developing an analytical model that allows for the incorporation of longer-term aspects of sustainability, we can greatly improve prospects for the future of agriculture in the tropics.

AGROECOLOGY AND ECONOMICS IN AGRICULTURAL DEVELOPMENT

Agroecology is the application of ecological concepts and principles to the design and management of sustainable agricultural systems. It has as a goal the development of agroecosystems that depend on low, external inputs; function more on the use of locally available and renewable resources; have benign effects on the environment; and are built upon the knowledge and culture of the local inhabitants of the region. The agroecosystem is the basic unit of analysis of any particular production system and, to a certain extent, can be analyzed using systems methodology common in most economic analysis.

Flow charts give considerable information for predicting the economic gain that can be expected from a crop. Materials and labor are inputs that have predictable costs, and harvest outputs can be assigned an income value. Depreciation can cover the replacement costs of farm equipment, and the maintenance or even eventual loss of certain elements of the farm system infrastructure can be discounted against crop yields (e.g., loss of an irrigation project because of silting-in of the reservoir from upstream erosion caused by deforestation) with a figure calculated on the basis of future replacement costs adjusted by inflation. The reduction of input costs most often has been achieved by reducing labor inputs, and by increasing outputs with material inputs such as fuel, fertilizers, machines, and pesticides. The costs of these inputs, though, have increased dramatically since the oil crisis of 1973, threatening the ability of increasing yields (and hence potential income) to continue to offset further augmentation of production costs without significant increases in food costs for the consumer. Broadly speaking, such an economic model views agriculture as basically a through-flow system. Inputs are added at one place or time so as to maintain or increase harvest removal, with losses being replaced the next season with further inputs. Management success is usually calculated on the basis of the margin of profit derived from fairly straightforward benefit/cost analysis.

When the agricultural system is viewed as an ecosystem (hence, the agroecosystem), we can begin to see that the through-flow production model has severe limitations. Material flow through an ecosystem is only a small part of what contributes to the long-term stability of that system. Inputs to natural ecosystems come from a broad array of biological and physical sources, and function less by moving through the system and much more by being retained and recycled within the system. Such inputs are usually limited in quantity, but they are quite constant over time. The living and nonliving components of the ecosystem are inextricably linked by complex sets of mutualistic, competitive, and often coevolved characteristics that lead to an ecological balance that, from an agricultural perspective, is enviable. By using such a closed or looped flow model for agroecosystems, we can begin to broaden the short-term, mainly economic focus of food production to include the longer-term aspects of ecological stability and sustainability. Productivity is viewed as a process, an understanding of which is tempered by aspects of nutrient cycling, energy flow, and the dynamics of plant and animal populations in the cropping systems (especially the crop and its pest organisms). Agriculture then has a basis for shifting toward a greater dependence on inputs from renewable sources, as well as using management strategies that are resource-conserving, environmentally sound, and contribute to the long-term productivity of agricultural lands. This also can contribute greatly to the long-term stability of the people and their cultures that so greatly depend on agriculture in tropical regions.

AN AGROECOLOGICAL APPROACH

Two aspects of an agroecological approach to agriculture in the humid tropics are necessary to ensure the development of information that is applicable, viable, and acceptable to the farmers or farming region involved. The first element is the knowledge and practical experience that local farmers have gained from season to season and generation to generation. Such knowledge reflects experience gained from the past, yet it continues to develop in the present as the ecological and cultural environment of the people involved goes through a continual process of adaptation and change. This has led to the development of agroecosystems that can make maximum use of locally available resources rather than rely so heavily on costly inputs imported from distant sources. They can simultaneously meet local needs together with supplying demands on a larger scale. A very important aspect, though, is the focus these systems have of maintaining productivity in the long-term, rather than an exclusive focus on the maximization of yields. The ability of these systems to keep the land permanently productive reduces the need for imported inputs, as well as the need for the development of new lands. Since such systems have evolved through time more as adaptations to local environments and factors, their study is beginning to provide us with information that contributes to the development of ecologically sound management practices that are understandable and acceptable to rural peoples in tropical regions.

It is upon this locally based framework that the application of agroecology can have major effect. By considering traditional or local resource management systems as part of a larger ecosystem, the impacts of any particular management practice can be understood. If enough information is available, these impacts can be, at least, partially predicted, helping to avoid some of the long-term environmental damage that has occurred in most tropical agricultural regions, as well as aiding in the establishment of more self-reliant farm systems. With an agroecological approach, then, human actions on the system can be placed in a context such as how nutrients enter, leave, or are recycled in the system, and ultimately aid in reducing the need for purchased fertilizers. Detailed study of how crop and noncrop plants and animals respond to different cropping arrangements or practices can continue to expand the ecological basis that has developed in recent years for the management of crop pests and diseases with much less dependence on costly and oftentimes environmentally damaging pesticides. A thorough knowledge of the interaction between physical factors of the environment (soil, water, air), their effect on, as well as modification by, biological factors (soil microorganisms, earthworms, herbivores), and the incorporation of the management of natural resources for human use, can permit the establishment of an understanding of both the agroecological potentials and limitations of the region under study. In additional, traditional or indigenous agroecosystems seem to reflect the more complex structure and function of tropical ecosystems in general. By examining a variety of tropical agroecosystems, we give a valuable means of beginning to test these ideas.

AN AGROECOLOGICAL RESEARCH FOCUS: CASE STUDIES

Annual Multiple-Crop Agroecosystems in the Tropics

The use of crop associations or multiple cropping can take place in many agricultural systems to great ecological and agronomic advantage. This is especially true in the humid tropics, where it is common to observe more than one crop occupying the same piece of land either simultaneously or in some type of rotational sequence during the season. Intensive agroecological studies of such mixed cropping have begun to provide a valuable basis for developing agroecosystems that are much less dependent on purchased inputs. At the same time production on the same piece of land can be intensified, resources can be more efficiently used and the land can be kept in more continual use.

One of the most common polyculture agroecosystems, especially in the Latin American tropics, is the traditional corn, beans, and squash system. In studies in Tabasco, Mexico, corn yield could be stimulated as much as 50 percent beyond monoculture yields when planted with beans and squash. Yield reductions for the two associated species were actually compensated for when the yields of all three were summed, giving higher total yields for the mixture as compared with an equivalent amount of land planted to the crops separately (overyielding). On the one hand, beans in corn polyculture appear to nodulate more and be more active in nitrogen fixation. This increased nitrogen input to the system may actually be available to the corn; this idea is supported by observations of net gains in nitrogen in the agroecosystem biomass despite the removal of harvest yields. Biological replacement of nitrogen reduces the need for a purchased input, and at the same time, some nitrogen may be actively transferred to the corn during actual association in the field through the use of mycorrhizal connections in the rhizosphere. Research is currently under way on this topic. Another aspect is weed control. The squash, rather than being planted for fruit yields, according to local farmers, was planted in order to control weeds. Thick, broad, horizontal squash leaves cast a dense shade that blocks sunlight, and combines with leaf washings that potentially inhibit weed growth through allelopathic interference. Finally, certain pests might be less of a problem in an intercrop system. The greater availability of microsites for shelter or reproduction of beneficial insects, as well as a greater diversity of food sources, may place the damaging insects at a disadvantage. A polyculture crop may have higher populations of biological control agents or, through greater variation in the structural diversity of the crop itself, it may be more difficult for the pest to find its host plant. Every component of the agroecosystem appears to play a role in maintaining productivity, the combined effects of which have been selected by farmers over many generations. Study of such systems can provide much useful information of value for the farmer.

INTEGRATED AGROECOSYSTEMS. The advantages of fully integrated agroecosystems, where plants and animals form part of the production system, have been much discussed recently. Annual and perennial crop plants are intercropped, herbs and trees combined, and animals can be free-ranging or confined. Even more important, production can be for both subsistence and market. Therefore, integrated systems have more of the different structural and functional elements common to natural ecosystems in the tropics, helping establish at least the ecological components of stable systems. The agroecosystem that seems to combine these characteristics the most is the tropical home garden.

The home garden is an integrated ecosystem of humans, plants, animals, soils, and water. Variously defined as kitchen, orchard, mixed, or forest gardens, the term "home garden" appears to be the most broadly used to describe a piece of land with definite boundaries usually near a house, occupying an area generally between .5 and 2 hectares. They are rich in plant species, usually dominated by woody perennials, and are generally described as being stratified or multistoried. A mixture of annuals and perennials of different heights forms layers of vegetation that resemble a natural forest structure. The high diversity of species permits year-round harvesting of food products, as well as a wide range of other products used by the inhabitants such as firewood, medicinal plants, spices, and ornamentals.

In a study of home gardens in both upland and lowland sites in Mexico, quite small areas (between .3 and .7 hectare) of high diversity were found to permit the maintenance of gardens that in some respects were similar to the local natural ecosystems. For example, the gardens studied had relatively high indices of diversity for cropping systems (Table 1) and had leaf-area indices and cover levels that approximated the much more complex natural ecosystems of the surrounding regions. In one study, nine different tropical ecosystems were analyzed for a series of ecosystem characteristics. A 40-year-old home garden had the most evenly distributed canopy that was fairly uniformly stratified from ground level to over 14 meters in height. Its leaf-area index was 3.9; cover was 100 percent; and the leaf biomass was 307 grams per square meter, the next highest for all of the ecosystems examined. Total root biomass down to a depth of 25 centimeters was identical to leaf biomass (307.2 grams per square meter). Most important, of the nine systems tested, the home garden had the highest root-surface area of small diameter roots (less than 5 centimeters) with a value of 2.1 square meters per square meter of ground surface down to 25 centimeters deep. These traits indicate an ecologically more efficient system, especially in its ability to capture light, garner nutrients in the upper layers of the soil, and store nutrients in the aboveground biomass, at the same time reducing the impact of rain and sun on the soil. Space between layers may increase carbon dioxide availability for photosynthesis.

Layers themselves may increase habitat diversity for birds and insects useful for maintaining better biological control in the system. Nutrients can be efficiently recycled through both ecological diversity and human manipulation.

TABLE 1

SPECIES TYPES AND CHARACTERISTICS OF HOME
GARDEN AGROECOSYSTEMS*

Characteristics	Cupilco	Tepeyanco
Average garden size	0.70 ha	0.34 ha
No. of useful species per garden	55	33
Diversity (bits)	3.84	2.43
Leaf area index	4.5	3.2
Percent cover	96.7	85.3
Percent light transmission	21.5	30.5
Perennial species (%)	52.3	24.5
Tree species (%)	30.7	12.3
Ornamental plants (%)	7.0	9.0
Medicinal plants (%)	2.0	2.8

SOURCE: Allison, 1983.
* Upland (Tepeyanco, Tlaxcala—four gardens) and lowland (Cupilco, Tabasco—three gardens) sites in Mexico.

An important aspect common in home gardens is the multiple uses that the various garden elements have. A cash crop such as coconut can also serve as a subsistence food. Some parts are used as fuel, while others for construction. Different plants or animals serve as important sources of carbohydrates, protein, vitamins, and minerals. For example, the fruit of the mango tree provides an excellent source of vitamins C and A. The mixture of species always offers something ready to be harvested, adding diversity to the diet. Many products can be sold for cash or traded. Variability in flowering and maturity ensures harvest, or even sale for income, during the entire year. Home garden advantages, then, have ecological components on the one hand in terms of management and maintenance able to rely on resource-conserving and low-external input practices. On the other, the diversity and flexibility of the gardens give the farmer assurance of constant, although variable, harvests.

AN AGROECOLOGICAL APPROACH TO WEED MANAGEMENT. Agroecology is not only concerned with the design and productivity of tropical agroecosystems. It is also very much focused on gaining a broader understanding of the management of factors that limit production. One of these factors of greatest importance, especially in the tropics, is interference from weeds. The modern approach to weed control has taken advantage of a broad array of technologies, the most widespread being the use of synthetically produced herbicides. In an agriculture where weeds are perceived only as something that lowers crop production, herbicides have proven very useful. But in agroecosystems that incorporate ecological elements into their management, a much broader conceptualization of the role of weeds is needed.

Weeds are important elements of the agroecosystem. Ecologically, weeds are well adapted to colonizing disturbed sites, enabling the establishment of soil cover for erosion control, slowing nutrient loss, raising levels of organic matter in the soil, improving nitrogen availability, conserving moisture, as well as possibly providing shelter and food for beneficial organisms and enriching the genetic pool of whole areas. In agroecosystems in developing countries where access to imported inputs is often very limited, many of these characteristics can be seen as having value in low-external input systems. With the inherent difficulty, then, of controlling weeds, there are definite advantages to not trying to work contrary to what are strong ecological characteristics and, instead, looking for ways to take advantage of them.

In the tropical lowlands of southeastern Mexico, local farmers were observed to not completely clear all weeds from many of their cropping systems. When questioned about these weeds (the Spanish translation is *malezas*), farmers mentioned that this is not the word normally used to refer to such noncrop plants. They use a concept of good and bad plants (translated as *buen* and *mal monte*)—a very utilitarian approach on how to consider these plants. Extensive interviews with farmers led to the idea that a special management of noncrop components exists within the agroecosystems of the region (Chacon and Gliessman, 1982). The recognition that weeds have both useful and detrimental characteristics has led to the establishment of management practices that not only work, but, as research is beginning to show, also have a strong grounding in ecological knowledge.

Close examination of weed management in traditional agroecosystems indicates that many tropical farmers possess a greater understanding of their systems than they are normally given credit for. Although they do not describe practices or problems in ecological terms, the relationship between some kind of vernacular classification scheme for weeds and ecological concepts can be established through appropriate research. For example, some weedy plants are classified as both good and bad by local farmers in Tabasco, Mexico. In tests designed to demonstrate the potential that these plants might have for inhibiting crops by releasing toxic chemicals (a factor known as allelopathy), great variability was found. A species such as *Lagascea mollis* is often cited by farmers for its ability to control noxious weeds as well as to serve as a feed for animals. It displays no significant allelopathic potential against corn and beans, the most important subsistence crops in the region, and is classified as a good plant and left in these cropping systems when encountered. *Guazuma ulmifolia*, a shrub found in fields soon after abandonment, actually shows stimulatory effects on corn. If it occurs in a field during cropping, it is frequently cut back to several centimeters above the ground, rather than eliminated, because of its use as an animal feed and the belief that it improves the soil and controls other weeds. On the other hand, a plant classified as good, like *Cleome spinosa*, displays very strong allelopathic potential in laboratory tests, but is found only as widely scattered individuals in most cropping systems and has a widely accepted use as a medicinal plant. It would be tolerated in the cropping system despite allelopathic potential. Some of those plants classified as bad by the farmer, such as *Paspalum conjugatum*, display rather limited allelopathic potential, but they compact the soil and cause the yellowing of associated crops. Part of the interaction could be competition for water or soil nutrients. Those plants classified as both good and bad could express allelopathic potential and thus be considered harmful for the crop, but good because of some other property, such as for medicinal purposes (e.g., *Chenopodium ambrosioides*).

Several leguminous noncrop species are intercropped with corn in Tabasco, Mexico. Two in particular have been shown to experimentally repress weed numbers and biomass when planted as cover-crop species. The results in Table 2 indicate that allelopathic potential may be one of the components of dominance over associated weeds. The fact that both legumes inhibited germination and growth of cabbage seeds and, though not as significantly, those of corn and beans, points to the possibility of tolerance in the crop components that have coevolved through time with weedy noncrop species. Cabbage is a relatively recent introduction to New World agriculture.

TABLE 2

GERMINATION AND INITIAL ROOT GROWTH OF SEED OF
THREE CROP SPECIES PLANTED IN WATER EXTRACTS OF LEAVES OF
TWO WEEDY LEGUMES IN TABASCO, MEXICO.***

| | Crop Test Species | | | | | |
| | Corn | | Bean | | Cabbage | |
Extract	% Germ.	% Growth	% Germ.	% Growth	% Germ.	% Growth
Canavalia ensiformis	98.9	77.6**	100.0	66.3**	49.3**	35.6**
Stizolobium deerengianum	100.0	87.9*	98.7	92.9	46.7**	27.2**
Distilled water control	100.0	100.0	100.0	100.0	100.0	100.0

SOURCE: After Gliessman, 1983.

NOTE: Numbers represent the means of ten seeds repeated three times (N = 30). Seed measured 72 hours following planting. Germination a simple percentage, with growth being measured of the main initial radicle, expressed as a percentage of control values.

* T test significant at 5%

** T test significant at 1%.

*** Expressed as a percentage of controls.

One of the legumes, *Stizolobium deerengianum,* is planted in rotation with corn and can effectively shorten the fallow time from four or five years to only six months. Allelopathic weed control, biological nitrogen fixation, and high organic matter inputs probably all combine to improve cropping conditions. It has been used in rotation with corn on some plots continuously for more than 20 years, maintaining yields at close to 3 tons per hectare dry grain without the need for additional nitrogen fertilizer or weeding (other than the initial ground preparation with machete at the time of corn planting). *Canavalia ensiformis,* although not often managed in rotation with corn, because of its bushy rather than vining habit, is associated as an intercropped species, thus aiding more directly in weed control during the actual cropping season.

The seed of *Stizolobium* is toasted and used as a substitute for coffee, while the young pods or mature seeds of *Canavalia* can be cooked and eaten.

In studies where crop yields are reduced by interference from weeds, competition is usually implicated as the reason. This is especially true in regard to nutrient availability. With their rapid growth rates and aggressive habit, weeds can capture nutrients that otherwise might be needed by the crop species. In a study of nutrient distribution in a traditional corn/bean cropping system in Tabasco, Mexico, it was found that weeds growing with the crops took up a very small percentage of the nitrogen taken up in the total crop system biomass (Table 3). Not until late in the fallow period, just before the clearing of the land for the next planting, did nitrogen in the weed component of the biomass reach its highest level. Light weeding with a machete during the growing season of the crops appears to reduce any possibility for at least nitrogen competition,

TABLE 3

NITROGEN DISTRIBUTION IN A CORN (*ZEA MAYS*)/BEAN
(*PHASEOLUS VULGARIS*) INTERCROPPED SYSTEM IN TABASCO, MEXICO.
1978 GROWING SEASON

| Month | N Content | | | | % N in |
	Crop Total	Litter	Weeds	Total	Weeds
January	9.86	133.	5.16	148.	3.5
February	53.1	114.	1.86	169.	1.1
March	91.3	96.7	5.60	194.	2.9
April	69.6	73.0	14.6	157.	9.3
May	0	121.	5.78	127.	4.6
June	0	135.	14.1	149.	9.5
July	0	52.8	0*	52.8	0
August	0	77.7	5.60	83.3	6.7
September	0	86.1	6.51	92.6	7.0
October	0	88.5	10.1	98.6	10.2
November	0	89.1	8.11	97.2	8.3
December	0	79.2	30.2	109.	27.7

SOURCE: After Gliessman, 1982.

* The field was cleared with machetes just prior to sampling; N = kilograms of nitrogen per hectare except as noted.

but the nutrient-capturing ability of the weeds play a very important role in trapping nitrogen in the fallow period.

An ecologically based approach to managing weeds is nothing new in traditional peasant agroecosystems. But it is in these systems, where weeds, crops, and humans have coevolved over a long period of time, that much information on how to manage weeds as integral parts of the agroecosystem can be gained. Studies of the ecological basis for management of weeds in their systems may provide agriculture of the future with a much greater degree of integrated management capability. This is especially important for farmers in developing countries where imported and costly inputs, such as most herbicides, only further increase the problems inherent in overdependence on externally generated technologies.

FUTURE RESEARCH NEEDS

An agroecological focus on tropical agriculture goes much beyond crop yields, and delves deeply into the complex set of factors that make up the agroecosystem. Local, indigenous agroecosystems that have evolved under the diverse and often limiting conditions of the tropical environment are adapted to this set of factors. They have evolved through time as low external-input systems, with a greater reliance on renewable resources and ecologically based management strategies. A research focus in agriculture that can take advantage of this knowledge and experience can permit us to explore the multiple bases upon which sustainability rests. It represents the blending of knowledge gained by ecologists who study the dynamics and stability of natural tropical ecosystems with the knowledge of farmers and agronomists on how to manage the complexities of food-producing agroecosystems. From this can come the sustainability in the production base so critical for the long-term cropping of farmlands in the tropics. Therefore, it becomes critical that we find ways to combine the short-term demand that economic models place on agricultural development with the need to project the long-term ecological models of sustainability generations into the future. We can do this only by demonstrating the absolute need to link ecological knowledge with long-term economic viability and environmental quality.

References Cited and Sources of Further Information

Alcorn, J.B., 1984. Huastec Mayan ethnobotany. Austin: University of Texas Press.

Allen, P., and D. Van Dusen (eds.), 1987. Global perspectives on agroecology and sustainable agriculture. Santa Cruz: University of California.

Altieri, M.A., 1986. The significance of diversity in the maintenance of the sustainability of traditional agroecosystems. Newsletter, Information Centre for Low External Input Agriculture (Netherlands) 3(2):3–7.

Altieri, M.A., 1987. Agroecology: the scientific basis of alternative agriculture. Boulder, CO: Westview Press.

Altieri, M.A., and M.K. Anderson, 1986. An ecological basis for the development of alternative agricultural systems for small farmers in the Third World. Am. J. Alternative Agric. 1(1):30–38.

Altieri, M.A., and D.K. Letourneau, 1984. Vegetation diversity and pest outbreaks. Crit. Rev. Plant Sci. 2:131–169.

Boucher, D.H., 1979. La nodulación del frijol en policultivo: el efecto de la distancia entre las plantas de frijol y maiz. Agric. Trop. (CSAT) 1:276–283.

Chacon, J.C., and S.R. Gliessman, 1982. Use of the "non-weed" concept in traditional tropical agroecosystems of south-eastern Mexico. Agro-Ecosystems 8:1–11.

Christanty, L., O. Abdoellah, G. Marten, and J. Iskander, 1986. Traditional agroforestry in West Java: the pekarangan (home garden) and kebun-talun (annual perennial rotation) cropping systems. *In:* G. Marten (ed.), Traditional agriculture in Southeast Asia. Boulder, CO: Westview Press, pp. 132–156.

Cox, G.W., and M.D. Atkins, 1979. Agricultural ecology. New York: Freeman, 397 pp.

Dewey, K.G., 1979. Agricultural development: impact on diet and nutrition. Ecol. Food Nutr. 8:247–253.

Douglass, G.K., 1984. Agricultural sustainability in a changing world order. Boulder: Westview Press, 282 pp.

Edens, T.C., C. Fridgen, and S.L. Battlefield (eds.), 1985. Sustainable agriculture and integrated farming systems. East Lansing: Michigan State University Press, 344 pp.

Ewel, J., S.R. Gliessman, et al., 1982. Leaf area, light transmission, roots and leaf damage in nine tropical plant communities. Agro-Ecosystems 7:173–185.

Gliessman, S.R., 1982. Nitrogen distribution in several traditional agroecosystems in the humid tropical lowlands of south-eastern Mexico. Plant and Soil 67:105–117.

Gliessman, S.R., 1983. Allelopathic interactions in crop-weed mixtures: applications for weed management. J. Chem. Ecol. 9:991–999.

Gliessman, S.R., 1985. An agroecological approach for researching sustainable agroecosystems. *In:* H. Vogtmann, E. Boehncke, and I. Fricke (eds.), The importance of biological agriculture in a world of diminishing resources. Witzenhausen: Verlagsgruppe, pp. 184–199.

Gliessman, S.R., and R. Garcia, 1979. The use of some tropical legumes in accelerating the recovery of productivity of soils in the lowland humid tropics of Mexico. *In:* Tropical legumes: resources for the future. Washington, DC: National Academy of Sciences, pp. 292–293.

Gliessman, S.R., et al., 1981. The ecological basis for the application of traditional agricultural technology in the management of tropical agroecosystems. Agro-Ecosystems 7:173–185.

Gonzalez Jacome, A., 1985. Home gardens in central Mexico. *In:* I.S. Farrington (ed.), Prehistoric intensive agriculture in the tropics. Oxford, England: BAR International Series 232.

Jackson, W., W. Berry, and B. Colman (eds.), 1984. Meeting the expectations of the land: essays in sustainable agriculture and stewardship. Berkeley: North Point Press, 250 pp.

Lappe, F.M., and J. Collins, 1982. World hunger: 10 myths. San Francisco: Institute for Food and Development Policy, 78 pp.

Lowrance, R., B.R. Stinner, and G.J. House, 1984. Agricultural ecosystems: unifying concepts. New York: Wiley, 257 pp.

Risch, S., 1980. The population dynamics of several herbivorous beetles in a tropical agroecosystem: the effect of intercropping corn, beans and squash in Costa Rica. J. Appl. Ecol. 17:593–612.

Vandermeer, J., S.R. Gliessman, et al., 1983. Overyielding in a corn-cowpea system in southern Mexico. Biol. Agric. Hort. 1:83–96.

Zimdahl, R.L., 1980. Weed-crop competition: a review. Corvallis, OR: International Plant Protection Center, 195 pp.

Chapter 2

ECOLOGY APPLIED TO AGROFORESTRY IN THE HUMID TROPICS

FLORENCIA MONTAGNINI

In the lowland humid tropics, more and more often, we see the need to develop alternative land-use systems that contribute to supply timber, fuel wood, and other tree products without continuing the well-documented patterns of extensive agriculture, deforestation, and land resources degradation. Alternative land-use systems should be productive while at the same time help preserve natural resources. Agroforestry systems can contribute to these goals when properly implemented.

Agroforestry systems, which combine trees with crops or pastures, are one of the alternatives for sustainable land use in the tropics. The incorporation of trees in the production system can shade animals and crops; improve the nutrient status of the soil, especially when nitrogen-fixing species are used; control soil erosion; provide tree products such as timber, fuelwood, fruits, and forage; make better use of the land through the association of species; and diversify the system, thus decreasing economic risks of crop failure and dependence on markets.

Proper choice of tree species for agroforestry systems can decrease the need for fertilizers and pesticides. For example, nitrogen-fixing species can help maintain soil fertility. In agroforestry systems, the appropriate association of trees with crops can reduce the need for herbicides, and insecticides. Costa Ricans, for example, grow coffee under open sun in extensive, high-input plantations owned by large companies. Small farmers cannot afford to buy the agrochemicals required. But they can grow coffee in association with nitrogen-fixing trees and other species valuable for lumber, fruit, and other tree products. The coffee plantations are productive for many more years and provide benefits beyond their cash-crop function. At the same time the risk of crop failure and market dependence decreases when the crop is diversified.

CHOICE OF TREE SPECIES

In agroforestry systems soil-improving tree species help increase productivity of low-resource agriculture. The improved supply of forest and agricultural products decreases the pressure over natural forests and virgin lands. Using fewer agrochemicals diminishes soil and water pollution. Better protection of headwaters decreases soil erosion and helps maintain water supply for agricultural, industrial, and domestic use.

Since the 1950s more people have become interested in experimenting with agroforestry practices. Some of these practices have existed for centuries, but are not well understood from a scientific point of view. In the neotropics several agroforestry systems are integrated into current practices and others are being developed. But, in most cases, studies that can help us understand mechanisms that influence their performance are missing.

In the Latin American tropics, studies are being conducted to test the performance of trees for reforestation or for agroforestry (Cozzo, 1976; CATIE, 1986; and others). Exotic species such as *Leucaena leucocephala, Acacia mangium, Gmelina arborea, Eucalyptus* spp., and *Pinus* spp. are widely used in agroforestry, although these represent only a small fraction of the trees that might prove useful in the tropics (NAS, 1983). Researchers concentrate on a small number of exotic species because of the experience generated in years of trials, genetic improvement, and silvicultural experiments.

However, the incorporation of native tree species offers several advantages. Native trees are well adapted to the local environment and are thus less likely to be affected by pests, diseases, and adverse weather conditions. Because they are adapted to local soils and vegetation, the design of mixed systems (as in agroforestry) is more feasible. Native species are in better balance with the natural ecosystem and allow better preservation of habitats for wildlife. Hence they help preserve biological diversity. The spread of the use of native species in profitable systems will increase the appreciation for these resources and help preserve the native flora of potential economic value.

In Costa Rica, tree trials for forestry and agroforestry have included exotic and a few native, generally fast-growing species. These trials are being conducted by institutions such as the Forest Service (Dirección General Forestal, DGF), Centro Agronómico Tropical de Investigación y Enseñanza (CATIE) (see, for example, Alpízar et al., 1986, CATIE, 1986; Glover and Beer, 1986), and private enterprises. In most of the regions where these projects are being conducted, we need strategies to help restore soil fertility, to decrease soil erosion and compaction, and to protect headwaters. However, very few projects devote resources to studies of potential effects of these tree species on soil properties (Montagnini, in press).

Research projects are needed to help fill this gap. In some cases there are not even enough data on general soil properties in the established systems. Thus, very often we cannot interpret differences in performance between systems. As a general consequence, it is often risky to extrapolate the results to other areas. Failure or success remains related to local conditions, and years and resources spent in the trials are not as profitable as they could be.

The Organization for Tropical Studies (OTS) and the Costa Rican Forest Service (Dirección General Forestal, DGF) have been studying the performance of 14 native tree species since 1985 at the OTS La Selva Biological Station (at Puerto Viejo de Sarapiquí, in the Atlantic Lowlands of Costa Rica) with funding from the Canadian Embassy (Table 1). The original objectives of the project were to provide local farmers with information on the potential of native trees for plantation forestry and agroforestry by establishing demonstration plots for use in a community-based environmental education program. Currently, funding is being requested from the U.S. Agency for International Development (AID) for an independent project to use the native tree-species plots to measure effects of these species on soil properties. The results of this project will provide more information on the native trees and increase the value of the demonstration plots.

TABLE 1

SPECIES USED IN THE OTS-DGF NATIVE TREES TRIALS

Tree Species	Family	Common Name
Hieronyma oblonga	Euphorbiaceae	Pilón
Calophyllum brasiliensis	Guttiferae	Cedro María
Cordia alliodora	Boraginaceae	Laurel
Dipteryx panamensis	Papilionaceae	Almendro
Lonchocarpus velutinus	Papilionaceae	Chaperno
Vitex cooperi	Verbenaceae	Cacho de venado
Dalbergia tucurensis	Papilionaceae	Granadillo
Pterocarpus officianalis	Papilionaceae	Sangrillo
Schizolobium parahyloum	Caesalpiniaceae	Gallinazo
Stryphnodendron excelsum	Mimosaceae	Vainillo
Tabebuia rosea	Bignoniaceae	Roble de sabana
Terminalia spp.	Combretaceae	Surá
Vochysia ferruginea	Vochysiaceae	Botarrama
Vochysia hondurensis	Vochysiaceae	Chancho

* Species of better growth after one year.

In January 1987, OTS began a major project funded by the Mac-Arthur Foundation to study the performance of exotic and native trees at La Selva, Costa Rica. It emphasizes growth performance of a large number of species; separate blocks were established for future use in studying the effects of 10 to 12 species on soil properties. Independent investigators can use the plots to study soil parameters and specific processes in far more detail than is planned in the MacArthur-funded project. Soil studies emphasize native tree species with the specific goal of putting existing regional resources into use for tree plantation and agroforestry.

A major contribution of OTS and CATIE to this area of applied ecology has been the production of the first agroforestry textbook in Spanish (OTS and CATIE, 1986), with funding from AID's Forestry Support Program. The book compiles information on the functioning of agroforestry systems, along with several case studies, bibliographies, inventories, and sources of information, covering most technical aspects. Besides its wide use as a university textbook and by extensionists, it represents a first step in gathering essential information useful as a starting point for a research program.

Soil-improving tree species are the key to the success of agroforestry systems. The performance of promising tree species and their effects on soil properties should be investigated in detail. This basic knowledge is needed before recommending the concept to farmers.

EFFECTS OF TREES ON SOIL PROPERTIES

The most important beneficial effects of trees on soil properties include improvement of soil structure and increase of soil nutrient availability. Deleterious effects such as decreases in soil acidity (pH) and soil nutrient content can also occur. Note, however, that this is a highly controversial issue (Cozzo, 1976; de las Salas and Fassbender, 1984; Lundgren, 1978; c.f. Sánchez et al., 1985) and that most of the reports of soil differences for different tree species do not indicate the processes involved (Lundgren, 1978).

Sánchez et al. (1985) concluded that the overall effect of mature fast-growing trees on nutrient availability is positive, but major differences exist among species. We also need to compare well-characterized soils, and to experiment with various tree/crop combinations compared with annual crops, pastures, or fallows to demonstrate to what degree trees improve or maintain soil properties in the humid tropics.

Tree species can influence soil pH, cations (calcium, magnesium, potassium), organic matter, and total nitrogen and phosphorus content and availability. For example, in a plantation of *Gmelina arborea* on an Ultisol soil in Brazil the soil pH was 5.2, while under *Pinus* (pine) the soil pH was 3.9—about the same as under native rain forest. The soil under *Gmelina* also had more exchangeable calcium (860 kilograms per hectare) than under pine (100) or native forest (40). Exchangeable manganese, exchangeable potassium, and total nitrogen under *Gmelina* were similar to rain forest, while the soil under pine had much lower potassium, magnesium, and total nitrogen content. Thirteen-year-old *Gmelina* plantations on Alfisol soil in Nigeria showed similar effects: an increase in soil pH from 4 to 5.5, a decrease in aluminum saturation from 58 percent to 6 percent, a two to three times increase in exchangeable calcium, and an increase in exchangeable manganese. Ten-year-old plantations of *Cordia trichotoma* and *Caesalpina equinata* showed a two to three times increase in exchangeable calcium, magnesium, and potassium, in comparison with the native forest; in contrast, no changes in soil properties were observed under *Dalbergia nigra*. These examples show how the effects of trees on soils differ among species.

Three species of eucalyptus exert the deleterious effect of extracting three to eight times more calcium and potassium from the soil than other forestry species, depending on silvicultural management. On the other hand, organic matter content was four times higher, and calcium, magnesium, and potassium content two times higher than in an adjacent pasture. Eucalyptus can absorb nutrients from deep layers of the soil with its 2- to 4-meter-deep root system, and contribute to increased soil nutrient content through litter-fall.

Controversial effects have also been reported for pine. Soils under an eight-year-old pine forest in Brazil had lower pH (4.2) and higher organic matter (5.3 percent) than an adjacent pasture of *Melinis minutiflora* 4.4 pH and 4.3 percent organic matter content. Soils under pine also had less calcium and much less potassium but higher magnesium content; there were no differences in phosphorus.

Apparently significant improvement of soil physical conditions—mainly increased porosity and decreased bulk density—occur only after tree canopy closure. Positive effects on soil physical properties are relevant in degraded lands where cattle have compacted soils, a common case in abandoned pastures in the humid tropics.

Obviously an understanding of the mechanisms whereby a tree species affects soil structure and nutrient availability is important to assess a species' true effects and to predict the outcome of its use. We must then use this understanding when designing agroforestry systems. In particular, knowledge of the effects of certain trees on soil properties can be used to achieve any management goals and to associate complementary tree/tree or tree/crop species.

THE ROLE OF APPLIED ECOLOGY: CASE STUDIES

Applied ecological projects can play a significant role by providing information needed for the design of successful systems. Studies of nutrient-cycling processes can be useful in the design and management of agroforestry systems, as the following examples illustrate.

Litter-fall (along with root decomposition) is frequently a main pathway of nutrient transfer (or recycling) from trees to soils. A thick layer of litter on the floor can also help maintain soil humidity and protect against soil erosion. These are also significant benefits from trees used for shade in agroforestry systems. For example, the net harvest output of nitrogen in a cacao plantation in Venezuela equaled the input of nitrogen to the soil in shade-tree leaf litter.

In Costa Rica, when organic matter and nutrients in the components of a system of cacao under shade trees were inventoried, leaf litter under the nitrogen-fixing leguminous tree *Erythrina poeppigiana* (commonly used in agroforestry systems in Costa Rica) was found to have more nitrogen than under *Cordia alliodora,* but less potassium, calcium, magnesium, and phosphorus. In another project in Costa Rica, the amount of nutrients recycled via litter-fall by associated trees (*Cordia alliodora* and *Erythrina poeppigiana*) was found to reach the recommended levels of fertilizer for coffee production.

Nitrogen-fixing trees are a prominent component of several agroforestry systems worldwide, because they can improve the performance of the associated crops or pastures. For example, nitrogen fixation by the leguminous tree *Inga jinicuil* was 53 percent of the average amount of fertilizer nitrogen applied annually on coffee plantations in Veracruz, Mexico. In Turrialba, Costa Rica, *Erythrina poeppigiana,* a widely used tree in agroforestry systems, was found to have a high potential for nitrogen fixation.

The influence of nitrogen-fixing tree species on soil nutrients other than nitrogen has also been reported, although the tendency has been to focus studies on effects on soil nitrogen and organic matter content. Soil properties in areas close and distant from *Erythrina glauca* trees in an old association with cacao in Brazil were compared: in three of five cases the pH was .2 to .3 units lower—closer to *Erythrina* than away from *Erythrina.* Phosphorus content was 50 percent lower under *Erythrina,* while calcium, magnesium, and potassium were twice as high under *Erythrina* than away from *Erythrina.* Soil bulk density was 1.55 under *Erythrina* and 1.61 away from this tree. It is interesting to notice the potential impacts of this nitrogen-fixing tree on phosphorus and soil pH.

Researchers in Costa Rica compared soils of cacao plantations associated with *Cordia alliodora* and with *Erythrina poeppigiana* trees. Soils under cacao plus *Erythrina* had 6 percent less phosphorus, 15 percent more organic matter, 22 percent more calcium, 15 percent more magnesium, and 19 percent more potassium than under *Cordia.* The effects of nitrogen-fixing species on soil nitrogen and cations can be positive, but questions arise as to their effects on soil pH and phosphorus content. We need comparisons of soil cations, phosphorus, nitrogen, organic matter, and pH between soils under nitrogen-fixing and other tree species to assess more realistically the role of nitrogen fixing and other tree species on these soil parameters.

Other projects stress the need of applied ecology to contribute to the design of agroforestry systems. In a hypothetical model of nitrogen, phosphorus, and potassium for a cacao/coconut association in Brazil, leaching of potassium was approximately 50 percent less than in a coconut monoculture. The appropriate choice of species for an association can thus contribute to nutrient conservation.

In southeast Bahia, Brazil, failure to expand promising agroforestry practices to a desirable extent was ascribed to the lack of scientific knowledge on the subject. As part of a crop diversification program in the traditional cacao-growing area, farmers combined clove trees with coffee or coconuts, cloves with passion fruit, papaya and coffee, cloves and vanilla, cardamom and coffee, and rubber trees, cacao, and other groups. This was successful, but the practitioners of these systems would not easily divulge the information on the management and production particulars in their farms. We need much more research and development if we are to realize the full potential of these systems.

PROMOTION OF APPLIED ECOLOGICAL PROJECTS

Applied ecological projects of the kind listed above may not always appeal to scientists who are interested in asking very specific questions and doing more sophisticated hypothesis testing. Since so much basic information is missing, many of these applied studies must start with preliminary inventories and comparisons. However, so many mechanisms of these systems need elucidation that it is hard to believe that scientists would not be stimulated to examine them. The implementation of applied ecology projects can be encouraged by science administrators who promote the funding of projects that would not be considered under their regular programs; by university professors and researchers from other academic institutions who encourage graduate students and fellow scientists to do theses and projects in the tropics; and by the scientific community as a whole, which can bridge the gap between basic and applied ecology and thus contribute to stimulating the kind of projects that are needed for this purpose.

Agroforestry systems—the combination of trees with crops or pastures—can be a productive and environmentally sound alternative land-use in the humid tropics. The choice of appropriate tree species is key to the success of these systems. Tree species differ widely in their effects on soils and in their nutrient requirements. The effects of tree species (including nitrogen-fixing trees) on soil properties remains controversial. We very much need projects that examine the performance of native tree species in agroforestry combinations, the effects of different tree species on soil properties, and tree/crop combinations that contribute to efficient nutrient usage and soil nutrient conservation. Applied ecology projects can contribute basic information on the systems to support the recommendations given to farmers.

References Cited and Sources of Further Information

Alpízar, L., H.W. Fassbender, J. Heuveldop, H. Folster and G. Enríquez, 1986. Modelling agroforestry systems of cacao (*Theobroma cacao*) with laurel (*Cordia alliodora*) and poro (*Erythrina poeppigiana*) in Costa Rica. I. Inventory of organic matter and nutrients. Agroforestry Syst. 4:175–189.

Alvim, R., and P.K.R. Nair, 1986. Combination of cacao with other plantation crops: an agroforestry system in Southeast Bahia, Brazil. Agroforestry Syst. 4:3–15.

Aranguren, J., G. Escalante, and R. Herrera, 1982. Nitrogen cycle of tropical perennial crops under shade trees. II. Cacao. Plant and Soil 67:259–269.

Cadima Zeballos, A., and P. de T. Alvin, 1967. Influencia del árbol de sombra *Erythrina glauca* sobre algunos factores edafológicos relacionados con la producción del cacaotero. Turrialba 17:330–336.

Centro Agronómico Tropical de Investigación y Enseñanza (CATIE), 1986. Silvicultura de species promisorias para producción de leña en América Central. Departamento de Recursos Naturales Renovables. Turrialba, Costa Rica, 219 pp.

Cozzo, D., 1976. Tecnología de la forestación en Argentina y América Latina. Buenos Aires: Ed. Hemisferio Sur, 610 pp.

de Barros, N.F., and R.M. Brandi, 1975. Influencia de trés espécies forestaes sobre a fertilidade de solo de pastagem en Viscosa, M. G. Brasil florestal (Rio de Janeiro) 6:24.

de las Salas, G., and H. Fassbender, 1984. Factores edáficos en los sistemas de producción agroforestales. *In* Heuveldop, J. and Lagemann, J. (eds.),

Agroforestería. Actas del seminario realizado en el CATIE, 23 febrero—3 de marzo 1981. Serie técnica—Boletín Técnico No. 14. Centro Agronómico Tropical de Investigación y Enseñanza. Turrialba, Costa Rica: Departamento de Recursos Naturales Renovables, pp. 30–36.

Fassbender, H.W., 1984. Bases edafológicas de los sistemas de producción agroforestales, Serie Materiales de Enseñanza No. 21. Turrialba, Costa Rica: Centro Agronómico Tropical de Investigación y Enseñanza (CATIE).

Glover, N., and J. Beer, 1986. Nutrient cycling in two traditional Central American agroforestry systems. Agroforestry Syst. 4:77–87.

International Council for Research in Agroforestry (ICRAF), 1983. Resources for agroforestry diagnosis and design, Working Paper No. 7. Nairobi: ICRAF, 383 pp.

Khanna, P.K., and P.K.R. Nair, 1980. Evaluation of fertilizer practices for coconuts under pure and mixed cropping systems in the west coast of India based on a systems analysis approach. *In:* K.T. Joseph, (ed.), Proc. Conf. Classification and Management of Tropical Soils, 1977. Kuala Lumpur: Malaysian Soc. Soil Sci., pp. 457–466.

Linblad, P., and R. Russo, 1986. C_2H_2-reduction by *Erythrina poeppigiana* in a Costa Rican coffee plantation. Agroforestry Syst. 4:33–37.

Lundgren, B., 1978. Soil conditions and nutrient cycling under natural and plantation forests in Tanzanian highlands, Reports in Forest Ecology and Forest Soils No. 31. Uppsala: Swedish University of Agricultural Sciences.

Montagnini, F., in press. Ciclaje de nutrientes en sistemas agroforestales. Primer Congreso Forestal Nacional, 1986. San José, Costa Rica.

Nair, P.K.R., 1984. Soil productivity aspects of agroforestry. Nairobi: International Council for Research in Agroforestry, 85 pp.

National Academy of Sciences (NAS), 1983. Calliandra: a versatile small tree for the humid tropics. Washington, DC: NAS, 52 pp.

Organización para Estudios Tropicales (OTS) and Centro Agronómico Tropical de Investigación y Enseñanza (CATIE), 1986. Sistemas Agroforestales. Principios y Aplicaciones en los Trópicos. San José, Costa Rica: OTS/CATIE, 818 pp.

Roskoski, J.P., 1982. Nitrogen fixation in a Mexican coffee plantation. Plant and Soil 67:283–291.

Sánchez, P.A., C.A. Palson, C.B. Dabey, L.T. Szott, and C.E. Russell, 1985. Tree crops as soil improvers in the humid tropics? *In:* M.G.R. Cannell and J.E. Jackson (eds.), Attributes of trees as crop plants. Huntingdon, England: Abbots Ripton, pp. 327–350.

Silva, L.F., 1983. Influencia de cultivos e sistemas de manejo nas modificacoes edáficas dos oxisols de tabuleiro (Haplorthox) do sul da Bahía. Belem (Brazil): CEPLAC, Departamento Especial da Amazonia.

Chapter 3

ECOLOGICAL ANALYSIS OF NATURAL FOREST MANAGEMENT IN THE HUMID TROPICS

ROBERT J. BUSCHBACHER

Between 1981 and 1985, an estimated 7.5 million hectares of closed tropical forest were clearcut each year (i.e., 0.62 percent of the total land area of such forests), and 4.4 million hectares per year were selectively logged.

Usually forests are clearcut to convert them to pasture or agriculture. Only a small proportion, if any, of the wood is harvested; most of it is simply burned. For example, in Honduras it was estimated that timber with a commercial value of 320 million U.S. dollars is annually destroyed in this manner. Conversion of forests to other land uses often results from large government subsidies that benefit wealthy corporations or individuals. In many cases, such as cattle ranching in the Amazon, the land uses being encouraged have proven ecologically *and* economically unsustainable.

Selective logging usually involves extraction of one or a few commercially valuable tree species such as mahogany in Latin America, *Khaya* in

Africa, or *Shorea* in Asia. Typically, only the largest, best-formed individuals are harvested for saw timber, plywood, or direct export as logs. If performed carefully, selective logging can inflict minimal damage on ecosystem structure and function. Nutrient depletion due to removal of a few boles per hectare is minimal and soil compaction due to logging roads and log skidding is usually localized. Indeed, selective cutting forms the basis for most of the natural forest-management systems discussed in this paper.

However, as it is currently practiced in most areas, selective cutting causes much more severe damage than the ideal. Severe impacts occur when heavy machinery is carelessly used, when steep slopes are exploited without proper erosion control, and when exploitation is so intensive that not enough seed trees of valuable species are preserved to ensure regeneration.

For example, increasing damage to poles and saplings in Malaysia has been attributed to the trend toward use of heavier equipment. In Sabah, Malaysia, 45 to 74 percent of residual trees are damaged or destroyed by harvesting operations. Erosion caused by harvesting steep slopes can adversely affect off-site areas, be they agriculture, water catchments, or even the Panama Canal. Finally, highly selective logging that focuses on very few valuable species and seeks to extract all desirable individuals can lead to the commercial extinction of those species from wide areas. An example of this severe impact occurred in the Amazon *varzea* (flooded forest) with *Virola surinamensis* and *Manilkara huberi*.

Selective logging is often performed carelessly, without regard for future productivity, because there is no economic incentive for sustainable management, and government control of logging activities is ineffective. Logging concessions are usually of short duration, and the concessionaire is unlikely to ever harvest a second rotation from a given area. Governmental agencies charged with monitoring and supervising logging activities are notoriously ineffective. Small staffs responsible for huge areas, lack of operational funds, and low salaries leave staff vulnerable to corruption. All contribute to the inability of these agencies to enforce rules to minimize environmental damage and ensure sustainability of the forest resource.

In addition to the direct impacts of logging, logging activities tend to make remote forest areas accessible, subjecting these areas to further disturbance that is often more severe than the logging itself. Landless peasants frequently follow logging operations, completing the deforestation by clearing agricultural plots. In addition to actual land invasion, selectively logged forest may be subjected to illegal harvesting of construction poles, mine props, or railroad ties. Finally, selectively logged forests are much more susceptible to damaging fires than undisturbed forests. These may spread from surrounding deforested areas being burned for agriculture or ranching or may be accidental.

The results of this widespread pattern of forest exploitation and destruction are both environmentally and economically costly. Habitat destruction is leading to species extinction and loss of nontimber forest benefits, such as erosion control, climatic buffering, and harvests of fuelwood, game, building materials, and other forest products. While some of these costs are felt on a global scale, it is local populations that bear the brunt of this burden, which translates into impoverishment, social upheaval, and migration to urban areas.

In addition, insofar as forest exploitation is not sustainable, commodity supplies will inevitably decline and forests will thus make a smaller contribution to the national economies of deforested countries. For example, Nigeria was once a substantial exporter of tropical timber, but must now import 160 million U.S. dollars' worth of wood annually. Indeed, Africa as a whole was a major log exporter in the two decades after World War II, but the forests were rapidly exhausted. In the 1960s, the focal point of international trade in tropical timber shifted to Asia, but this resource may soon be depleted as well.

Three major alternatives to existing patterns of tropical forest exploitation could produce forest products while maintaining long-term ecosystem productivity: plantation forestry, agroforestry, and natural forest management. This chapter does not attempt to evaluate the relative costs and benefits of these three approaches to tropical forestry. It assumes, rather, that each approach is valuable under the appropriate circumstances. The remainder of the chapter seeks to evaluate the natural forest-management systems that have been attempted in various tropical areas.

EXPERIENCE WITH NATURAL FOREST MANAGEMENT

Schmidt (1987) has defined natural forest management as "controlled and regulated harvesting, combined with silvicultural and protective measures, to sustain or increase the commercial value of subsequent stands, all relying on natural regeneration of native species." Compared with plantations and agroforestry, natural forest-management systems are less intensive; they have relatively low yields but demand smaller capital input and require less productive land. Natural forest-management systems are much more similar to natural forest than either of the other alternatives.

There has been a great deal of experience with natural forest-management systems in Asia and Africa. The many management systems that have been tried include shelterwood systems, which promote regeneration by partial canopy opening; polycyclic felling systems, designed to mimic the age and size distribution of natural forest; and strip clearcutting followed by natural regeneration.

Malayan Uniform System

Perhaps the classic example of a natural forest-management system is the Malayan uniform system (MUS), developed shortly after World War II for lowland dipterocarp forests rich in *Shorea* species. The MUS developed in response to the growing mechanization of log extraction and milling processes, which provided incentive for single, heavy exploitation operations; and to the observation that areas exploited without any silvicultural treatment during the Japanese occupation had adequate spontaneous regeneration.

In brief, the silvicultural operations of the MUS consist of harvesting all merchantable trees in a single operation, followed by the poison girdling of all unwanted larger stems. About three to five years after exploitation, diagnostic sampling is conducted to determine the status of the regeneration and prescribe silvicultural treatments to ensure vigorous and healthy regeneration. Silvicultural treatments to promote regrowth include climber cutting and poison girdling of weed trees, defective stems, and stems that compete with trees of the most valuable species. This intervention is repeated at 10, 20, 40, and 60 years, and the next cut occurs after 70 years. The goal of this suite of treatments is to obtain an even-aged stand of predominantly dipterocarp species.

Evaluations of the MUS over two decades of operation indicate that regeneration was satisfactory on many of the sites. However, while silviculturally satisfactory, the fatal flaw of the MUS was its dependence on the same areas of even topography and fertile soils that were most in demand for agriculture by the burgeoning population of Malaysia. As a result, the focus of forestry in Malaysia today has moved away from the lands where the MUS was designed to function, onto the hillslope forests. On these sites, advance regeneration is too sparse and too heterogenous to ensure adequate stocking after harvesting, and the MUS has been abandoned.

The difficulty of transferring the silvicultural techniques that worked so well in Malaya's lowland dipterocarp forests to the surrounding hillslope forests illustrates the limitations on transferability of the MUS to other regions. In short, the MUS is only silviculturally appropriate when two preconditions are met: an adequate stocking of seedlings of desirable species prior to harvesting; and a large enough proportion of commercially valuable species in the original forest canopy to justify complete canopy removal.

Tropical Shelterwood System

The tropical shelterwood system (TSS) was introduced in Nigeria in 1944 and Ghana in 1946. In contrast to the Malayan uniform system, the TSS could not rely on abundant existing regeneration. Since the potential seed trees were not very abundant and seeded at irregular intervals, it was necessary to begin canopy opening several years prior to harvesting in order to promote adequate regeneration. Canopy opening by felling or poison girdling of selected trees was prescribed over a five-year pre-harvest period.

In practice, almost every tree not considered economic was killed, resulting in drastic canopy opening (65 to 80 percent of basal area removed) and removal of several species that were later considered commercially valuable. Indiscriminate canopy opening could not attain the delicate regulation of forest-floor light levels needed to favor regeneration over weeds and vines. The result was that climber tangles often spread "like a cancerous growth under the canopy openings."

In sum, the TSS was considered unsuccessful because of low regeneration rates by the prime economic species, climber infestation, and the high labor costs associated with the numerous, intensive interventions. In addition to these silvicultural limitations, the TSS in Nigeria was superseded by intensified land uses such as oil palm and cocoa plantations, while exploitative forest logging for export accelerated.

Polycyclic Felling Systems

In response to the perceived inadequacy of the shelterwood systems described above, the modified selection system was developed in Ghana in the 1950s, as was the selective management system in Malaysia in the early 1970s. These systems emphasize advanced regeneration, i.e., residual trees below the minimum cutting limit at the time of initial harvesting, to comprise the subsequent crop on a 25- to 30-year cutting cycle.

A key silvicultural consideration for design of such polycyclic felling systems is the length of the cutting cycle. In both Malaysia and Ghana, short-term economic requirements had greater influence on determination of the length of the cutting cycle than did silvicultural feasibility. If growth rates prove inadequate to meet minimum size criteria by the end of the cutting cycle, there are three possible responses: increase minimum cutting limits in order to leave more advanced regeneration, extend the cutting cycle, or increase growth rates through liberation thinnings.

The first response, increasing minimum cutting limits, selects out the fastest growing individuals and results in "creaming," or genetic depletion of the stand. This threatens the long-term sustainability of logging practices.

Increasing the length of the cutting cycle is a more conservative response to insufficient growth rates by residual trees because it ensures long-term genetic integrity of the stand. The exact length of the cutting cycle in a given stand should not be fixed by any universal prescription; the management system should be flexible enough to respond to occasional site inventories that monitor stand development. Sophisticated stand-development models derived from well-designed surveys and executed on microcomputers can make management much more sensitive to the specific conditions on a given site.

Liberation thinning is a silvicultural treatment designed to enhance the growth of economically valuable understory trees by killing neighboring competitors. This treatment is most appropriate for secondary or recently logged forests that have an open canopy. Selection is accomplished by dividing the stand into quadrats 10 or 15 meters square and selecting the most promising individual on the basis of species (commercial value), size, and form. Competing stems to be removed are chosen based on their size and proximity to the target stem, as well as canopy overlap.

Liberation thinning removes 15 to 20 percent of the total number of trees, concentrated in small patches, or wells, around the selected stems. Liberation thinning counteracts the progressive forest impoverishment that results from repeated selective logging, and has been characterized as indispensable to a minimum cutting limit.

Strip-Clearcut System

The Palcazu Rural Development Project in eastern Peru has been comprehensively designed to incorporate silvicultural, economic, and social considerations, making it one of the most exciting and promising experiments in natural forest management ever carried out in Latin America.

Silviculturally, the forest is being harvested by long, narrow clearcuts designed to mimic natural forest disturbances, such as tree-fall gaps, by the width of the strip, the lack of burning or heavy machinery, and reliance on natural regeneration. Two demonstration strips cut in 1985 showed that harvesting and extraction of wood can be accomplished under local conditions without gross environmental damage, and that initial regeneration is rapid and abundant. Critical aspects of the harvesting operation include proper selection and orientation of the strips, careful design of access roads to minimize erosion, and no felling of trees outside the strips to minimize disturbance to nonharvested forest.

The economic basis of the forest management plan is complete usage of all wood larger than 5 centimeters in diameter. Wood larger than 30 centimeters in diameter will be cut for saw timber, while smaller logs will be preserved for use as telephone poles and construction posts by high-pressure injection of heavy metal salts. Smaller wood will be converted to charcoal in portable kilns. All products will initially be marketed in the local region, and thus access to and price in local markets is crucial to project success.

Socially, the Palcazu Rural Development Project is based on collective management of communally owned lands by indigenous people. To a large extent, the long-term success of the forest-management plan will depend on the capability of these communities to manage the logging, processing, and marketing operations. For example, it is vital to maintain the sawmill and preservative system despite difficult environmental conditions and tenuous access to spare parts. In addition, it is important to schedule extraction activities carefully so poles and posts are preserved quickly after harvesting, and to avoid transport of heavy logs during high rainfall periods. Considerable management skill will also be necessary to ensure that transport is available at the proper time and, if necessary, to alter the product mix in response to market conditions.

Finally, appropriate distribution of project benefits among community members will be essential for project success. The Amuesha must perceive direct and continuous benefits from the forest resource, or they will burn and farm the clearcut areas. Indeed, burning and farming are central to the Amuesha cultural tradition. However, natural regeneration by desirable species following logging is dependent on minimal disturbance to the site, in particular requiring total absence of fire or cultivation. Thus, tight control over land-use management practices, especially after removal of the large initial economic subsidies, must be demonstrated before this project can be considered successful.

In summary, the Palcazu Rural Development Project is comprehensive in its conception and promising in its execution, but tight control of land use, sophisticated management of operations, and longer studies of the rate of natural regeneration are necessary before its success can be evaluated. This project will be closely watched in both the conservation and development communities as a possible model for environmentally sound rural development.

Low-Intensity Clearcutting

In the 60,000-hectare Bajo Calima forest concession in northwest Colombia, Carton de Colombia has been practicing careful, low-intensity clearcut harvesting to supply a nearby pulp mill. Trees are felled with axes or chainsaws, cut into convenient lengths and debarked in the field, and transported from the forest to a road by a system of skyline cables. These harvesting methods minimize forest perturbation in spite of the large volume of wood harvested. No heavy machinery is used anywhere other than the logging roads. Debarking and trimming branches before removal leaves the material with greatest nutrient concentrations on the site. The soil is left covered by both this residue and the intact litter layer.

The growth rate of the natural regeneration after low-intensity clearcutting was higher than for any of 13 species experimentally planted in small monoculture plots. Biomass after 15 years of natural regrowth was about half of the biomass before logging. Diversity and species composition of the natural regrowth was similar to the pre-logging forest.

These observations indicate that natural regeneration following careful, low-intensity clearcutting is silviculturally feasible at Bajo Calima for at least two rotations. To predict long-term sustainability, studies of nutrient losses from harvesting, leaching, and erosion as compared with input from soil weathering and the atmosphere, would have to be performed. Certain special conditions at Bajo Calima, furthermore, indicate limitations to transferability of this system. The key to minimizing site disturbance was the use of cable logging, which may not be feasible in a forest with trees of larger diameter than the relatively small stature forest at Bajo Calima.

The economic key to the project was integration with a company-owned pulp mill, where techniques had been developed for complete utilization of mixed tropical hardwoods. The company has produced kraft pulp and paper packaging from mixed tropical hardwoods since 1959, and bleached pulp for writing paper since 1983. These innovations, coupled with targeting domestic rather than the more demanding international market, are of extreme economic importance because they grant the flexibility to manage the native forest according to its limitations rather than replacing it with monoculture plantations of gmelina, pine, or eucalyptus.

The major limitation to the success of Carton de Colombia's land management results from social pressures, or incompatability with the surrounding population and economy. As regeneration has reached pole size four to six years after clearcutting, local people have entered the site via the old logging roads. They typically make heavy and repeated selective cuts for poles, mine props, and construction posts. High unemployment in the nearby city of Buenaventura and the proximity of major highways make this activity economically attractive to the local people. Thus, most of the regeneration is not allowed to return to mature forest and cannot be harvested in a second rotation. Company control efforts have been either lax or ineffective—even a major research plot was disturbed by pole cutters. The fact that unharvested portions of the concession that did not have logging roads were not invaded illustrates the role of accessibility in promoting chronic disturbance.

The only solution to this problem would be incorporating local communities in continued forest management throughout the regeneration period. This amounts to an intensification of the land use. A general conclusion, well illustrated by this case study, is that the extremely low-intensity land uses, often necessary to reduce costs in natural forest-management systems, may be socially inappropriate in densely populated regions. Land that has been made easily accessible and that is managed only passively is unlikely to remain free of chronic disturbance.

AN INTEGRATED APPROACH TO
FOREST MANAGEMENT

Natural forest-management systems have both conservation and economic benefits. Natural forest-management systems are economically sustainable because harvesting operations are combined with cultivation and provision for regeneration of the next rotation. Thus, harvesting should be repeatable into the future, assuming that a small enough area is logged from a given management unit each year, and that cutting cycles are long enough.

Ecologically, natural forest-management systems are sustainable because: soil compaction and erosion are confined to limited areas and minimized by careful log extraction; nutrient removal in harvested material is small relative to total ecosystem nutrient stocks and can be replenished by atmospheric inputs and mineral weathering; and utilization of a diverse array of native species reduces susceptibility to catastrophic outbreaks of pests or diseases.

However, forest-management systems cannot be successfully designed in a socioeconomic vacuum. In addition to silvicultural sustainability, forest management must be economically profitable and socially feasible, or it will fail. Neither an exclusively financial nor a strictly ecological approach will suffice. In a recent article destined to be a classic of forest economics, Leslie (1987) has demonstrated the inadequacy of financial accounting, with its emphasis on short-term cashflow, to fully analyze the costs and benefits of alternative forestry investments. The key gaps in traditional financial analysis are: failure to consider nonrevenue benefits and external effects; market imperfections and distortions in the actual prices of inputs and outputs; and an arbitrarily determined, overvalued discount rate.

On the other hand, a long list of silvicultural systems, apparently viable and sustainable from an ecological viewpoint, have been abandoned, invaded, destroyed, or otherwise failed. These examples debunk the notion that ecological sustainability is a sufficient condition to ensure rational forest management.

Just as ecological analysis routinely incorporates climate, topography, soils, and plant, animal, and decomposer communities, a truly holistic analysis must also incorporate people who are or may exploit the forest, markets that determine the limits of commercial viability, accessibility to processing facilities and consumers, and so on.

Climate

The most important climatic variables for forest management are annual precipitation and length of dry season. Low rainfall is often a constraint to tree productivity. This is particularly true during establishment on disturbed sites, where the number of successive dry days may be the critical determinant of seedling survival. On the other hand, high rainfall areas are subject to intensive soil disturbance and erosion during log extraction operations.

Topography

Slope is very frequently the chief limitation to land-use capability. Steep slopes make logging much more expensive and dramatically increase the amount of soil erosion that results. In addition, high elevation, steep areas are often hydrologically important, yielding water for agriculture, urban use, and hydroelectric installations. Nevertheless, such considerations may not prevent steep areas from being appropriate for well-managed, low-intensity natural forest management.

Soil Fertility

The fertility of tropical soils is often related to geological age. Ancient landforms are highly leached and tend to have low cation exchange capacity. Infertile, red-clay Oxisol and Ultisol soils are typical of vast areas of Amazonia, Africa, Borneo, Indochina, and other tropical regions. These soils are typically acidic and low in available phosphorus. In contrast to these ancient soils, volcanic, geologically young, and alluvial soils are often very fertile. These soils are common in Central America and on some of the islands of the Pacific and Asian regions. Soil fertility can have a clear effect on wood growth rates, but due to low turnover rates, undisturbed forests on infertile sites can also reach very large stature. Thus, tree size in undisturbed forest is not necessarily a good indicator of soil fertility.

Forest Structure

The total volume of wood on a site and its distribution among trees of various size-classes are important considerations for management. Smaller size-classes are most appropriate for construction poles, railroad ties, charcoal manufacture, and pulp. Larger size-classes have higher value for a given volume of wood because they can be more efficiently converted to saw timber or plywood. In terms of management, it may be profitable to delay harvesting beyond the age of maximum girth increment because of the greater production by larger trees of high value, dense heartwood.

Species and community characteristics are also important as determinants of the ease of management of a site. For both management and marketing reasons, forests dominated by one or a few species are excellent targets for management—the Dipterocarp forests of Malaysia are the classic example of this situation. Seasonally or permanently inundated forests are also often dominated by a single species; examples include the *Raffia* palm swamps of Latin America and the *Cativo (Prioria copaifera)* forests of Colombia and Panama.

In terms of species ecology, the key factor is the regeneration biology of the most valuable or dominant species. The easiest species to manage are gap-opportunist species that persist in low light conditions but are capable of rapid height growth in full light. Completely shade-tolerant species may not tolerate full light, and are not generally capable of rapid growth. Light-demanding species are fast growing, produce ubiquitous seed, and can dominate highly disturbed sites. These light-demanding species generally produce low-quality wood, have poor growth form, and are of low economic value. The majority of silviculturally valuable species are thus in the intermediate gap-opportunist category.

Marketability

The species present on a site have obvious importance in determining the value of the resource. In addition to log size and form, the chief characteristics that determine wood quality are density, workability, and resistance to decomposition. Dense woods are difficult to work; in addition, a high silicon content tends to dull saw blades. Resistance to decomposition can be increased by preservation with toxic salts, which is being tried in the Palcazu Rural Development Project. For cabinet and furniture woods, color and grain are also important, while the chief determinant of pulp value is fiber length. In addition to inherent quality, two additional factors that affect marketability are how well a species is known on the market and how uniform the wood of a given species or stand tends to be. Familiarity with wood characteristics facilitates processing, and uniformity simplifies mill management in terms of adjusting processing equipment and techniques. A large diversity of wood types requires different saws, different equipment settings, and so forth.

Far Eastern producers have had great success in grouping species of similar properties for sale as a single wood class. Asabere (1987) recommends that wood from Ghana be grouped into density classes when appearance is not critical and when availability does not justify promoting species separately. The volume of wood of a given type available and continuity of supply are key marketing factors.

Another important aspect of marketability is the potential for sales at a local level. For example, poles and posts do not have sufficient value to bear a large transport cost, but if they can be sold in local communities they could recover the costs of thinning treatments and provide income at an intermediate stage of the rotation. On both a micro and a macro scale, there has been a widespread historical process whereby the number of wood species that are economically valuable continuously increases. For example, the list of economic species in Nigeria increased from 17 to 40 between 1953 and 1961. In the Ivory Coast, trees other than the seven most valuable species increased from 15 to over 50 percent of the harvest between 1952 and 1978. This is a result of increasing accessibility over time, as well as a response to depletion of the most desirable species.

Previous Disturbance

Prior disturbance may have had a negative effect on forest structure and species composition. On the other hand, indigenous inhabitants may have enriched the forests with valuable tree species. For example, *Manilkara zapote* and *Brosimum alicastrum* are valuable multipurpose tree species which were of great importance to the Maya (most notably for production of abundant, edible fruits and a valuable gum, respectively) and are now abundant throughout the Yucatán. *Euterpe* palms, which produce a wide range of products including food, medicine, and construction materials, may have played a similar role in the Amazon.

In addition, previous exploitation may have built up an infrastructure, primarily in the form of roads, that could make future management much more cost-effective. For example, the Belize Estate Company property is an area that was heavily logged in past decades but has recently been unexploited. The old logging roads in this area could facilitate management of the present forest.

Local Population Pressure

Expanding agriculture in a region is an important factor affecting the viability of low-intensity, long-term land uses such as natural forest management. Invasion by squatters, illegal logging, and the spread of fires from agricultural or ranching areas often limit the success of forestry efforts.

Inhabitants who do not have clear and certain land title may be reluctant to make long-term investments in improving the land's productive capacity. Likewise, larger landholders or industrial firms are hesitant to invest in the long-term productive capacity of a site if they are uncertain that they will be able to maintain control of an area for the many decades that may be necessary to receive the anticipated income. Land-tenure instability pushes forest exploitation toward short-term mining activities and is one of the chief impediments to long-term management.

Transport

Since transport costs can be the key determinant of stumpage prices, accessibility is tremendously important in determining the viability of a proposed forest-management system. This includes both distance to market and the status of transport routes such as roads and rivers.

REGULATION, MARKETING, AND LAND CONTROL

The three keys to successful natural forest management, corresponding to silvicultural, economic, and social considerations, are regeneration, marketing, and control of the land. These three aspects are inextricably linked. For example, the amount of canopy opening prescribed by a given silvicultural system may or may not be practical depending on the marketability of the main forest species. In another case, pre-harvest treatments such as climber cutting or tree marking may not be practical if they necessitate opening the forest by road construction and if this would lead to illicit logging, cutting of railroad ties, or invasion by squatters.

Silvicultural Sustainability

The critical factor in the silvicultural analysis of natural forest management is the effect of harvesting and intermediate treatments on regeneration. Silvicultural research and experimentation should focus on the effects of canopy opening on light, moisture, and temperature of the forest floor environment, and how these factors in turn affect the interaction between regeneration of desirable species relative to competing weeds and vines.

The answer for any given case will depend on the ecology of the target tree species: its pattern of flowering and fruiting, seedling growth rates and environmental requirements, competitive ability, and so on. By definition, a desirable silvicultural treatment is one that opens the canopy enough to encourage rapid growth by the valuable regeneration, but does not open the canopy so much that weeds and vines can outcompete that regeneration.

Another essential prerequisite for effective silvicultural management is an efficient mechanism for conducting forest inventories. Natural forest-management systems are less capital-intensive, but more information-intensive, than traditional forest management. For optimal efficiency, it is necessary to maintain flexibility with respect to pre-harvest regeneration release, post-harvest weeding, and the length of the cutting cycle. Rather than rigid prescriptions, the timing and extent of these treatments should be based on periodic inventories of regeneration abundance and growth rates.

It is critical that the responsible forest-management agency, whether the Forest Service or a concessionaire, allocate sufficient resources so that all cost-effective interventions are indeed carried out. In the case of a government agency, this would probably require that a proportion of forest-harvesting revenues be funneled directly to the component of that agency responsible for field activities. In the case of a private concessionaire, ensuring conscientious forest management will require a carefully crafted set of government-monitored incentives and sanctions. One mechanism toward this end is requiring that logging operations begin with less accessible areas within a given concession; harvesting of easier, more profitable stands can be predicated on compliance with all management requirements.

Two final silvicultural considerations that may affect the long-term productive capacity of a stand are the effects of harvesting operations on site nutrient pools and soil structure. Direct nutrient removal results from harvesting of the boles. Since nutrients are more concentrated in the bark than in the wood itself, the amount of nutrient depletion could be significantly reduced by manually debarking logs before extracting them from the site; this is routinely performed at the Carton de Colombia Bajo Calima concession.

Soil disturbance due to logging operations is a function of slope, rainfall, and the size of equipment used in log extraction. Only the last of these three factors is under the direct control of the land manager. Using lighter equipment, in conjunction with smaller concession sizes and tighter control of logging practices, can decrease soil disturbance at the same time that the utilization efficiency of the forest stand is increased. These measures are critical to ensuring silvicultural sustainability of tropical forest management.

Economic Profitability

Probably the most critical prerequisite for improved management of tropical forests is an increase in the number of commercially valuable tree species. This need reflects back directly on silvicultural techniques, because it gives the forest manager effective control over the degree of canopy opening. In addition, increasing the number of economically valuable species increases the likelihood of sufficient regeneration on a site either before or after initial harvesting. Finally, utilization of more species would in effect result in the wood resource being more concentrated, necessitating fewer roads per cubic meter of wood harvested, and requiring a smaller radius of forest to supply a given processing plant.

In order to foster efficient use of their forest resources, tropical countries should impose policies that ensure that harvesting is not limited to one or two species. Consuming countries, too, should agree to criteria that guarantee that the wood they import is fairly and sustainably produced. The International Tropical Timber Organization is a potential venue for such agreements among producing and consuming countries.

Several countries have promoted marketability of secondary species by promoting a domestic home-building program that uses those woods. Research on wood properties and aggressive marketing and information campaigns are other components of a total strategy to increase wood marketability.

When natural forest management is applied in the context of rural development, income at intermediate stages of the cutting cycle becomes even more important than total return. In addition to intermediate harvests, the other important mechanism for making income more continuous over time is to manage several blocks of forest, each at a different stage in the production cycle. This is relatively straightforward for governmental or industrial forest managers. For rural communities, income continuity may only be achievable if individual producers form cooperatives to jointly manage a property much larger than any individual could afford to own.

Finally, development of local, small-scale processing industries has the potential to increase the rural income component of forest-harvesting activities. Such projects are important in their own right, but would have the added benefit of building a local constituency concerned with continued productivity of the forest resource. The economic self-interest of local inhabitants is a critical tool for maintaining the integrity of the forest resource.

Social Feasibility

One of the critical limitations that reduces the willingness of investors to make long-term investments in tropical forestry is uncertainty in the ability to maintain control over the land. This hesitancy is justified by the land-use history of forested tropical regions. Time and again, forest-management plans that seem ecologically sustainable and economically profitable have failed because of inability of the forest manager to control land use. A good example is the Carton de Colombia Bajo Calima concession, where natural regeneration after clearcutting was reported to be good and prospects of continued pulp supply to a local mill were promising. Due to the access granted by the roads built for initial harvesting, however, the regeneration was destroyed after five or so years of regrowth by unemployed local people cutting construction poles and mine props.

In Malaysia, the Malayan Uniform System functioned effectively on the lowland dipterocarp forests where it was developed, but land-use changes in the last decade or two have resulted in the land being converted to agriculture. Forestry operations have been moved to hill-slope forests where the Malaysian Uniform System was less suitable silviculturally.

The latter example illustrates the point that forestry operations are most appropriate on marginal lands. Promotion of forestry on land suitable for intensive agriculture can have unintended adverse consequences, if it displaces subsistence farmers and forces them onto unsuitably steep or infertile land. Current reforestation incentives in Costa Rica and India, for example, are mainly affecting large landowners on prime agricultural lands. As a result, the stated goal of reforestation of marginal lands is only being partially met, while unexpected adverse effects in the agricultural sector may occur.

Another lesson from the many failures of silviculturally sound forestry systems is the necessity to demonstrate active land occupation and utilization. Thus, a successful management strategy must be sufficiently intensive, relative to competing land uses, to avoid having partially harvested forest become an attractive target for land invaders. This is a real difficulty, because the economic keystone of natural forest-management systems is to compensate for a relatively low rate of return by minimizing management costs.

Given these social constraints on natural forest management, it may be necessary for governments to guarantee that long-term investments in forestry will not be lost by actual or de facto land expropriation by agricultural interests or incoming populations of subsistence farmers. Thus, the role of governments in assuming risk in forestry investments needs to be considered.

Three kinds of risks are associated with forest management: biological risks, such as pests and diseases; climatic risks, such as hurricanes; and social risks, such as land invasion, expropriation, or, in some cases, fire. One of the virtues of natural forest-management is that it minimizes biological risks. Climatic risks may be unavoidable, but they are somewhat mitigated by the relatively low capital investment required for natural forest management. However, if a government is committed to ensuring the sustainability of production while maintaining the secondary benefits of forest cover, it may be necessary to assume some of the social risks of natural forest-management investments.

Such insurance would follow the precedents of crop insurance and disaster relief that are widespread in developed economies. The costs of such insurance may be high, but they should be incorporated into the economic assessment of forestry investments, and may be found to be cost-effective. An added benefit of this "social-disaster insurance" is that it would provide a powerful incentive for governments to make and promote policies that stabilize rural land-use. This, in turn, would minimize the likelihood of the social disaster actually occurring.

References Cited and Sources of Further Information

Adriawan, E., and S. Moniaga, 1986. The burning of a tropical forest: fire in East Kalimantan. Ecologist 16(6):269–270.

Anderson, A., in press. Use and management of native forests dominated by acai palm (*Euterpe oleracea* Mart.) in the Amazon estuary. Adv. Econ. Bot.

Asabere, P., 1987. Attempts at sustained yield management in the tropical high forests of Ghana. *In:* F. Mergen and J.R. Vincent (eds.), Natural management of tropical moist forests: silvicultural and management prospects of sustained utilization. New Haven: School of Forestry and Environmental Studies, Yale University, pp. 47–70.

Baur, G.N., 1964. The ecological basis of rainforest management. Sydney: Forestry Commission of New South Wales, 498 pp.

Carton de Colombia, 1985. Ninth annual report, forest research in the Bajo Calima concession. Cali, Colombia: Carton de Colombia, S.A.

Fearnside, P.M., and J.M. Rankin, 1985. Jari revisited: changes and the outlook for sustainability in Amazonia's largest silvicultural estate. Interciencia 10(3):121–129.

Fox, J.E.D., 1968. Logging damage and the influence of climber cutting prior to logging in the lowland dipterocarp forest of Sabah. Malayan Forester 31(4):326–347.

Gillis, M., in press. Malaysia: public policies, resource management and the tropical forest in Sabah, peninsular Malyasia and Sarawak. *In:* Repetto, R. and M. Gillis (eds.), Public policy and the misuse of forest resources. Cambridge: Cambridge University Press.

Hecht, S.B., 1982. Cattle ranching development in the eastern Amazon: evaluation of a development strategy, Ph.D. dissertation. Berkeley: University of California, 454 pp.

Hutchinson, I.D., 1979. Liberation thinning: a tool in the management of mixed dipterocarp forest in Sarawak. *In:* Proceedings of the Seventh Pan-Malaysian Forestry Conference, Penang, Malyasia.

Instituto Nacional de Investigaciones sobre Recursos Bioticos (INIREB), 1983. El Chicozapote: informa: comunicado no. 59 sobre recursos bioticos potenciales del país. Xalapa, Ver.: INIREB.

Kunstader, P., 1978. Alternatives for development of the upland areas. *In:* P. Kunstadter, E.C. Chapman, and S. Sabhasri (eds.), Farmers in the forest. Honolulu: East-West Center.

Lanly, J.P., 1982. Tropical forest resources, Forestry Paper No. 30. Rome: Food and Agriculture Organisation. 106 pp.

Leonard, H.J., 1987. Natural resources and economic development in Central America. New Brunswick, NJ: Transaction Books, 279 pp.

Leslie, A.J., 1987. A second look at the economics of natural management systems in tropical mixed forests. Unasylva 155(39):46–58.

Lowe, R.G., 1968. Tropical shelterwood system investigations in the moist semi-deciduous forest of southern Nigeria. Enugu: 2nd Nigerian Forestry Conference, 1966.

Lowe, R.G., 1978. Experience with the tropical shelterwood system of regeneration in natural forest in Nigeria. Forest Ecol. Mgmt. 1:193–212.

Mergen, F., and J.R. Vincent (eds.), 1987. Natural management of tropical moist forests: silvicultural and management prospects of sustained utilization. New Haven: School of Forestry and Environmental Studies, Yale University.

Pardo-Tejeda, E., and C. Sanchez Munoz, 1981. *Brosimum alicastrum:* a potentially valuable tropical forest resource. Xalapa, Ver.: Instituto Nacional de Investigaciones sobre Recursos Bioticos.

Parker, G.G., 1986. Analysis and recommendations concerning the natural forest-management system used in the Central Selva project, Peru. Mill-brook, NY: New York Botanical Garden, 25 pp.

Rankin, J.M., 1985. Forestry in the Brazilian Amazon. *In:* Prance, G.T. and T.E. Lovejoy (eds.), Key environments: Amazonia. Oxford: Pergamon Press, pp. 369–392.

Schmidt, R., 1987. Tropical rain forest management: a status report. Unasylva 39:2–17.

Strugnell, E.J., 1947. Developments in silvicultural practice in Malaysian evergreen forests. Malaysian Forester 2:37–41.

Tropical Science Center, 1982. Sustained yield management of natural forests. San José, Costa Rica: Tropical Science Center.

Uhl, C., and R. Buschbacher, 1985. A disturbing synergism between cattle ranch burning practices and selective tree harvesting in the eastern Amazon. Biotropica 17(4):265–268.

van Schaik, C.P., and E. Mirmanto, 1985. Spatial variation in the structure and litterfall of a Sumatran rain forest. Biotropica 17(3):196–205.

Wadsworth, F.H., 1969. Posibiliadades futuras de los bosques del Paraguay. Asuncion, Paraguay: Ministerio de Agricultura y Ganaderia.

Walton, A.B., 1948. Some considerations for the future management and silvicultural treatment of Malayan forests. Malaysian Forester 2:68–74.

World Resources Institute, 1985. Tropical forests: a call for action. Washington, DC.

Chapter 4

INSECT ECOLOGY AND AGRICULTURAL PEST MANAGEMENT: THEORY AND PRACTICE

AGNES KISS

No discussion of applied ecology in the service of economic development can be complete without a segment on agricultural pest management. This is a field that epitomizes the application, or frequently the failure to apply, basic ecological principles to the task of meeting basic human needs. In the most basic sense, the relationship between the human species and the plethora of potential species of insect, weed, and microbial pests is a competition to exploit the niche of consuming agricultural products. In this rather uneven competition, our unique scientific and technological ingenuity is pitted against their sheer reproductive capacity.

Since synthetic organic pesticides were introduced in the 1940s, we have been relying increasingly on one rather simplistic tactic—eliminating the competition with chemical weapons. However, this approach is becoming less and less successful, due to the unwitting application of several other fundamental ecological and evolutionary principles. While this may be considered unfortunate from the point of view of agricultural economics, in another sense it may be viewed as a mixed blessing, as it lends persuasive economic arguments to support those who feel pesticide use should be significantly reduced worldwide because of growing environmental and health problems. If familiar pesticides continued to work as effectively now as they did in the 1940s and 1950s, agricultural producers would be less interested in attempts to limit their use.

Many economic entomologists claim that the number one constraint to successful chemical control of pest insects is the development of insecticide-resistant populations—an inevitable outcome of combining strong selective pressure and enormous reproductive potential. Insecticide resistance to sulfur compounds was first observed in the San Jose scale insect in 1914. As described by Metcalf (1984):

At the most recent tally in 1980, insecticide resistance was documented in 428 species of insects: 260 pests of agriculture and 168 pests of human and animal health. Multiple resistance of pest species, or insect resistance to several of the various chemical classes of insecticides, is a far more serious phenomenon. As of 1980, 2-stage resistance was documented in 105 species, 3-stage resistance in 64 species, and 4-stage resistance in 26 species. At least 14 major insect pests are now resistant to all 5 classes of insecticides (DDT, lindane/cyclodienes, organophosphates, carbamates, and pyrethroids).

As a further complication, the rate of resistance development to newly introduced classes of insecticides has shown an exponential increase. At the same time that development costs of new pesticides are increasing rapidly due to technical and legal factors, the anticipated useful life of these products is rapidly decreasing through development of cross-resistance in target pest populations, primarily among insects, but increasingly among weeds and pathogens as well. One familiar example is the rapid development of resistance to pyrethroids among populations resistant to DDT.

Another very significant constraint to successful chemical control is pest resurgence; that is, the rapid development of even larger populations of the original target species, or the emergence of new pest species, following a large-scale chemical control program, which may have been quite successful in the short term. Again, a fundamental ecological principle is responsible: a predator/prey equilibrium has been upset as many of the natural enemies have been destroyed along with the pest organisms upon which they would ordinarily feed. Predators are often more affected by chemical attack than are herbivores, both because they may be more active and therefore more exposed to air-borne sprays, and because of bioaccumulation of toxins in the bodies of the prey. The populations of the pest species recover more quickly—as might be predicted for primary versus secondary consumers—and then explode beyond their original bounds because the density-dependent regulating factors of predation and parasitism have been removed.

The common response to pest resurgence in the field has been to increase the dosage and frequency of pesticide application. This may or may not be effective in controlling the original target species, depending upon how readily resistance evolves in the population. What it almost always seems to lead to, however, is the emergence of new pests. These are usually species that were present in the area, and possibly even attacking the crop prior to the pesticide barrage, but were at such low densities that they were not regarded as particularly troublesome. As their competitors and the predators and parasites that held them in check are reduced, however, still another ecological principle—competitive release—comes into play and they increase to economically significant, and sometimes devastating, levels.

These secondary pest species are often insects whose feeding habits or life styles give them some immunity to chemical attack. Among the most familiar examples are the boll weevil, the bollworm, and the whitefly in cotton. These insects, which spend a large part of their life cycle sheltered inside cotton squares or on the undersides of leaves, have repeatedly emerged as secondary pests following widescale chemical control of more exposed pests, such as the lepidopterans *Heliothis* and *Spodoptera*. They have made cotton growing almost impossible or uneconomical in many areas of the world at one time or another (e.g., the southwestern United States, areas of Mexico, Peru, Brazil, and Sudan).

ECOLOGICAL PEST CONTROL (INTEGRATED PEST MANAGEMENT)

Pesticide resistance, pest resurgence, and the emergence of secondary pests are examples of what can happen when we attempt to undertake agricultural pest control while ignoring agricultural ecology. The technology of applying ecological and entomological science to pest control is commonly referred to as integrated pest management (IPM).

While the term IPM has been widely used and misused, it is best defined as the integrated use of all available methods to maintain pest populations below an acceptable threshold level. That level depends on a number of factors, including the type and extent of the damage caused by a particular type of pest, the economic value of the crop, and the cost of control measures. IPM tends to emphasize preventive measures such as planting resistant varieties, using crop rotation, improved crop hygiene, and protecting natural enemy populations, because these are more cost-effective and more likely to continue to work in the long term. Contrary to some popular misconceptions, IPM does not completely eschew the use of chemical pesticides; these are considered to be one weapon in the arsenal. However, they are to be used judiciously and conservatively because they are expensive, because they are self-limiting due to resistance, and because their use tends to reduce the effectiveness of other measures, particularly biological control. In popular parlance, IPM has sometimes been extended to include minimizing negative impacts on human health, wildlife, and other elements of the natural environment. However, while these are important goals, they are not inherent in the original concept of IPM.

It is important to emphasize that IPM is not so much a technology as it is an approach or even a philosophy. It is defined not by the various elements of biological control, resistance, cultural practices, and so forth, but by the commitment to integrate these in a comprehensive management plan that weighs costs against benefits and emphasizes stability and returns over the long term, rather than short-term maximization of yield or profit. This has important implications for the role of scientists in promoting the IPM approach, as it means that simply developing effective individual control methods is not enough. To qualify as IPM, these control methods must be combined into a rational system that is compatible with both local agroecology and socioeconomics.

Overall, IPM seeks to restore an ecological equilibrium of sorts, with humans included among the primary consumers (or secondary consumers in the case of livestock husbandry). Unfortunately, this is becoming increasingly difficult to do as the human element becomes increasingly demanding. With our own population growing rapidly (though not by entomological standards) and with the area of arable land also decreasing through urbanization, desertification, etc., agriculturalists are under continuous pressure to produce more and more calories from each plot of cropland. This means denser planting, year-round cropping, shortened crop rotations with less time in fallow, use of high-yielding varieties often planted as extensive monocultures, increased mechanization, and increased use of fertilizers. All of these factors favor the further expansion of pest populations and make it more difficult to identify compatible, ecologically sound methods of pest management. For example, it is already well recognized that increased use of nitrogen fertilizer is associated with increases in pest densities, particularly in the case of homopterans such as aphids and leafhoppers. There is now also evidence that use of some pesticides can affect the metabolism of plants in ways that make them more susceptible to pests and diseases (Chaboussou, 1986).

Undoubtedly for these reasons various entomologists over the years have gone on record with different variations of the same theme: when it comes to pest control, we may win occasional battles, but we are not winning the war; while we can control almost any specific pest (although sometimes at a prohibitive cost), the overall problem of controlling pests is more acute than ever.

APPLICATION OF INTEGRATED PEST MANAGEMENT EVALUATION OF PAST EFFORTS

The preceding review of the relationship between pest management problems and ecological principles contains no new revelations. We know and understand the basic ecological and evolutionary principles involved, and the outcomes are fairly readily predictable. However, the subject is not ecology per se, but the application of ecology to economic development, in this case the development of intensified agricultural production in the humid tropics. Thus three questions are important:

• To what extent is this knowledge being applied in the agricultural field, particularly in the humid tropics?
• If this is not happening to an appreciable extent, why not?
• What can we as scientists and development professionals do about it?

The response to the first question would have to be that the performance has been mixed. Certain aspects of the IPM approach have been spreading successfully in tropical regions. The most basic and ancient IPM technology is cultural control. This encompasses a wide range of activities including: crop rotations, which alternate susceptible and nonsusceptible crop species or varieties; the destruction of diseased plants to avoid the spread of contagious diseases; early planting to escape peak pest emergences; removal of crop residues to kill dormant stages of boring insects; manipulation of water levels in irrigated rice as a method of weed control, and so forth. One well-dated form of cultural pest control that has awakened renewed interest in recent years is intercropping or polyculture. While the evidence from research is mixed, indications are strong that, at least in some cases, crops grown in polyculture suffer less from pest and disease attack than do the same species and varieties grown in monoculture. Weed control can also be enhanced by polyculture, through shading out and, in some cases, allelopathy. While not a new approach, cultural control is continually being refined through new research and through the development of new crop varieties such as short-season cultivars.

The use of insect- and disease-resistant crop varieties is another fundamental component of IPM that is emphasized in plant breeding programs in many national and international research institutes, such as the International Rice Research Institute (IRRI) in the Philippines, the International Potato Center (CIP) in Peru, the International Institute for Tropical Agriculture (IITA) in Nigeria, and the International Center for Insect Physiology and Ecology (ICIPE) in Kenya. As a result of years of research at these and other sites, there are now, for example, widely available rice varieties resistant to pests such as the brown plant hopper and stem borers and a number of diseases; potato varieties resistant to the tuber moth, aphids, and mites; maize varieties resistant to stem borers and streak virus; cassava varieties resistant to the mosaic virus; okra-leaf cotton resistant to the whitefly. As new problems have arisen through parallel evolution of new strains of the pest species adapted to attack the resistant crop cultivars, current research focuses increasingly on development of moderate horizontal resistance to a range of pests rather than more complete vertical resistance to a single one.

Biological control is probably the technology most often associated in our minds with the IPM approach. Some very successful examples of introduction or augmentation of biological control agents include the introduction of the parasitic wasp *Epidinocarsis lopezi* to control the cassava mealybug in West Africa, baculovirus against rhinoceros beetle in coconuts in the South Pacific, and *Trichogramma* egg parasites against sugarcane borers in mainland China. Introduction of biological control agents is most often successful against introduced pest species, and augmentation is currently possible only with a relatively small number of parasites and predators that can be mass-reared. Another, and equally important, aspect of biological control is safeguarding indigenous populations of natural enemies. In the banana plantations of Central America, the cotton fields of the Canete Valley of Peru, the rice fields of Indonesia, and many other sites, growers have learned the hard way that careless elimination of local predators and parasites can lead to disastrous explosions of pest populations. Fortunately for world agriculture, in many cases the agroecosystem has proven sufficiently resilient that when spraying was halted or significantly reduced by decree, the populations of natural enemies returned and the pest species were brought under control once again.

In light of experiences such as these, movement toward more selective use of pesticides has been significant, in terms of reducing frequencies and volumes of application and total areas affected. One example is the control of tsetse flies in Africa. In the past, tsetse fly populations were reduced or eliminated over huge expanses of land through use of broad-scale, high-volume aerial sprays of persistent products such as dieldrin (along with massive forest and brush clearing and slaughter of game animals). More recently, systems have been developed that combine the use of screens soaked with attractant pheromones and pyrethroids, ultra-low volume, sequential spraying with endosulfan at low doses, and spot-spraying of resting sites in the undergrowth. The technology remains imperfect and significant kills of birds, fish, and beneficial insects still occur, but the long-term environmental impacts of chemical control are significantly reduced when these methods are used. (The environmental impacts associated with clearing tsetse flies from forest areas, thus permitting the entry of cattle is, of course, another question that will not be taken up here.) Unfortunately, these new methods require a greater level of technical expertise and logistical organization to implement, and very few skilled practitioners are available.

The increased cost of pesticides is, of course, an additional factor favoring conservative application and the development of very-low-volume and ultra-low-volume methods. Perhaps most telling have been attempts to reduce the use of broad-spectrum insecticides that take such a toll on natural enemies. The most notable case is in Indonesia, where the government recently banned the use of a number of such products in the rice fields, specifically on the basis of pest resurgence problems. In Indonesia, the officially recommended product for control of brown planthopper in irrigated rice is now the juvenile hormone analogue Buprofezin.

In addition to these examples of applications of individual key components of the IPM approach, there are a few examples of comprehensive IPM programs. These are defined here as an approach in which definitive control thresholds are applied and a range of chemical and nonchemical control strategies are integrated into a unified system. Most of these examples are from industrialized countries, notably the United States, where a nationwide IPM project has led to the adoption of the IPM approach in production of key crops such as cotton, soya, alfalfa, and fruit trees over large areas of the country. In the tropical regions, the best known examples of successful comprehensive IPM programs are probably those instituted for cotton in Peru and in Nicaragua, following earlier virtual elimination of cotton production due to problems with pest resistance and resurgence. However, views differ on how well these ambitious and initially successful programs have stood the passage of time and the prevalence of armed conflict. More recently, a national program promoting IPM in soybeans has been initiated in Brazil.

APPLICATION OF INTEGRATED PEST MANAGEMENT—CONSTRAINTS

Overall, however, reliance on chemical pesticides continues to grow in tropical countries, and much of this chemical control continues to follow in the old, destructive patterns of prophylactic and calendar spraying, high-volume applications, and use of broad-spectrum toxins. This leads to a consideration of the second major question: What is impeding the implementation of IPM, particularly in the developing countries of the tropics? If our objective is to act to enhance the spread of the IPM approach, it is particularly important to determine the relative importance of technological, economic, political, or sociological constraints.

In the view of many traditional crop-protection experts, the most important constraint is that IPM is by its nature too information-intensive and too complex for application outside a research station setting. In making this criticism, they are not necessarily referring to the computer programs that some U.S. farmers now use to analyze market information to make cost/benefit decisions, to predict pest population densities based on weather conditions and to calculate ideal combinations and dosages of pesticides. Even without approaching this level of sophisticated technology, IPM is a relatively complex proposition.

Chemical control can be effective (in the short term) even if practiced very simplistically. Spreading a blanket of potent biocides over a field is certain to kill a large portion of whatever animal life is present, including herbivorous insects. By contrast, under IPM the farmer seeks to target the particular species that causes significant damage. He or she must be able to identify at least the major potential pest species (and distinguish them from beneficial species), estimate their abundance as well as their probable impact on the final yield, and evaluate what that means in terms of potential total loss. Finally, pest and predator life histories must be sufficiently understood to know the most vulnerable point for attack and to avoid interfering with natural control mechanisms. In addition to this sophisticated application of biology, ecology, and economics, the farmer must be aware of the range of management options available, including the pros and cons of particular pesticides if chemical control is determined warranted. Most important, a range of control options that are feasible must be available. Cultural methods that require high input of labor at the same time that there are other important tasks to be done may not be feasible, nor are rotations with crops for which there is no market, nor burning crop residues needed for animal fodder or building materials.

Clearly, one does not expect individual farmers, particularly peasant farmers in developing countries, to possess all of this information and sophisticated knowledge. It is the role of agricultural researchers and extensionists to provide guidelines and rules of thumb to follow, such as economic thresholds for important crop/pest combinations and guidance on cultural practices and on pesticide selection and application methods to target specific pests. The question of how well this role is being filled, and whether it can be filled better, is addressed later.

The second potentially important constraint—site specificity—is related to the information factor. Not only does IPM demand a relatively large input of information, but that information cannot be generalized from one country, ecological zone, farming system, or growing season to another. Because IPM seeks to achieve an ecological balance, the dynamic nature of the agroecosystem is an important factor. The key pests, their impact on yields, and the potential for natural control are all variable from site to site and from year to year, and an IPM program must be very flexible to respond effectively. To draw upon an earlier example, the whitefly may be the most significant pest of cotton in Sudan, but it is virtually unknown in Central and South America. In Sudan, it is also a greater problem in wet than in dry years. Thus, while the theory of IPM can be taught, these details make it very difficult to transfer IPM technology from developed countries to developing countries; relevant research must be carried out on-site to develop suitable technology packages. This presents a particular problem in developing countries, where research is often poorly organized and almost always underfunded.

A third technical constraint that is significant in temperate countries, but perhaps more so in the tropics, is the presence of multiple pests. Unlike the use of broad-scale chemical control that can make a dent in the populations of a number of species simultaneously, a basic principle of IPM is targeting of specific pests. In temperate areas it is more likely to be feasible to target each important pest species, individually, with highly specific cultural practices and pesticides. In the tropics, biological diversity in general is much greater, so that the number of potential pests is larger. For example, significantly more diseases occur in the same crops in tropical versus temperate zones, as shown in Table 1.

TABLE 1

NUMBERS OF DISEASES REPORTED IN
CROPS IN TWO CLIMATES

	Temperate	Tropical
Citrus	50	248
Pumpkin squash	19	111
Sweet potato	15	187
Tomato	32	278
Rice	54	500–600
Beans	52	253–280
Potato	91	175
Maize	85	125

As an additional complication, measures that help contain one pest problem may aggravate another.

The existence of multiple pests makes it difficult to market highly specific pesticides even in the United States, where the chances are that a farmer can identify the problem and get what is needed when it is needed. In developing countries where available cash and effective distribution systems for input are far less common, farmers must try to prepare ahead of time for whatever may arise. In practice, this means a strong preference for relatively inexpensive, broad-spectrum pesticides with few specialized requirements for application methods or equipment.

A fourth constraint of particular significance in developing countries is sociopolitical rather than technical, relating to the distribution and control of land. For IPM to be most effective, it must be practiced in a coordinated fashion over large, contiguous areas. At the same time, it must be responsive to varying conditions within different portions of that area. This is possible on large farms in the United States, where a single owner or manager controls the whole area and has a direct stake in its productivity and efficiency, and where field information is relatively easy to collect and process rapidly. In developing countries, much of the agricultural land is held in the form of small parcels. If pest management is under the control of the individual smallholders, an effective area-wide IPM program would require considerable coordination and cooperation among them—an objective very difficult to achieve. If, as is the case in many African countries, the pest-control operations are run directly by the state, there is, in theory, a better opportunity for coordinated implementation. However, there is also often less possibility and incentive to respond to actual variable on-ground conditions.

The inevitable bureaucracy involved in state-run operations means that a farmer's plot will probably be sprayed when it has been scheduled to be, and not when scouting results indicate that it should be. Individual farmers could undertake nonchemical measures, but they may simply consider all pest control the responsibility of the government (often paid for automatically with taxes or obligatory fees or figured into the fixed prices of crops). Crop-protection service employees are usually not accountable to the farmers, making it hard to ensure quality of scouting or control operations.

Finally, on the socioeconomic/political side, there is the question of incentives for and against implementation of IPM. In the United States, the IPM approach has probably been spreading for three principle reasons: because chemical controls fail; because pest-control costs can often be reduced through IPM; and because of growing consumer and citizen movements concerned with reducing pesticide residues in foods and the environment.

In the humid tropics, chemical control is failing in many cases, and sometimes it is well documented that this is due to resistance problems, for example, failure of BHC or lindane to control mirid bugs in cocoa in large areas of West Africa. Very often, however, it is not clear why chemical control fails. Farmers often blame the ineffective performance of national crop-protection services, or adulterated pesticides, and they are undoubtedly often right. Consequently, they call for improved chemical control, rather than for an alternative approach. Just as important, it is often difficult to demonstrate to the farmers' satisfaction that IPM represents a viable, practical alternative. Where extension services are weak and pesticide salesmen represent the farmers' sole or primary source of pest-management advice this is a particular problem. Sales representatives usually work on a quota or a commission and are unlikely to promote a reduction in use of their products. Even modifications, such as using more specialized pesticides, are usually unpopular with farmers, at least at first, as they lack the dramatic effect of the large-scale kill to which the farmers have become accustomed through their use of highly toxic, broad-spectrum products.

Cost reduction is an important potential incentive; unfortunately, it often does not function in the context of developing country agriculture. Repetto (1985) argues that widespread government subsidization of pesticides in developing countries reduces the financial incentive of farmers to minimize their use or seek alternatives. Moreover, in Pakistan, elimination of these subsidies was followed by significant reductions in the total volume of pesticides used. It can be argued that developing-country agriculture represents a far from perfect market overall, and that pesticide subsidies may be needed to restore a balance in light of government control of crop prices. Nevertheless, the fact remains that subsidization of pesticides reduces the incentive to minimize pesticide use. A related problem is the distribution of low-cost or free pesticides by manufacturers seeking to build a market for their products, and even by bilateral assistance agencies as a form of commodity aid. A further factor working against IPM is the prevalence of state-run pest-control operations, whose costs are subsumed in crop prices, or supported by tax funds or other non-optional sources.

The third potential incentive, consumer concern about pesticide residues, comes into play in most developing countries, at present, primarily with regard to export crops. For example, several West African countries now ban the use of lindane on cocoa primarily because their major customers demand this. Similarly, the use of dieldrin on ground nuts and forage crops has been significantly reduced worldwide based on instances in which ground nut or beef export shipments were refused by European countries or the United States because of dieldrin residues. Local consumers in developing countries remain largely ignorant of the issue of residue problems, whether in food, drinking water, or the environment, and national governments have little or no relevant regulation or means of enforcement. In some countries, such as Thailand and India, this is beginning to change, largely due to the efforts of locally based nongovernmental organizations, with some support from the United Nations specialized agencies and other donors, including the World Bank. The most visible effect of concern over pesticide residues has probably been the widespread reduction in the use of DDT, and to a lesser extent, dieldrin, endrin, chlordane, and toxaphene, for agricultural purposes. Many tropical developing countries have followed the West in instituting such a ban for these few high-visibility products.

INCREASING THE APPLICATION OF IPM
OPPORTUNITIES FOR ECOLOGISTS

This brief summary of some important factors that impede the spread of IPM as a method of pest management in tropical countries leads to the final and most pressing area for discussion—what can scientists and concerned, yet foreign, citizens do to enhance the adoption of the IPM approach? The industrialized countries have a stake in promoting world stability through, among other things, environmental and agricultural stability. Improved agricultural pest management represents one area of economic development in which biologists and ecologists should be in a position to offer intellectual and technical leadership.

Consideration turns to research and teaching. Clearly there are important data gaps to be filled in order to develop appropriate and practical pest control methods and make IPM a viable approach in any given agroecosystem. Some key areas include:

- crop-loss assessment research to establish valid control thresholds;
- information on life histories and population dynamics of pest species, particularly the role of natural control agents;
- rearing and dispersal methods for natural enemies;
- socioeconomic factors in agricultural practices, to identify the realistic constraints and motivations that underlie farmers' responses to proposed new technologies.

However, the role of expatriate scientists in this research has changed over the past few decades. In recent years, international support for national research programs and for training of developing-country scientists has created a cadre of well-trained and motivated researchers, many of whom are strong proponents of IPM. These scientists are often employed in a national agricultural research service or in agricultural technical colleges or universities, and are both capable and motivated to carry out this type of basic research if they receive at least a minimum level of support. Development assistance agencies are becoming less interested in financing research programs initiated and directed by expatriates, preferring instead to support national research programs with relatively short-term participation of expatriates in an advisory capacity. When U.S. or other expatriate scientists become directly involved in on-site research projects, it is increasingly on the basis of collaborative programs recognized as being as much for the benefit of these researchers' professional interests as it is for the developing country. Their greatest contribution in some cases may be to help channel research funds into the local program; they can often also help to improve the effectiveness of locally conducted research by introducing new research methods, in particular, improved data-management methods often based on computer technology. In addition, training of developing-country researchers and technicians remains an important priority, although an increasing tendency is to provide that training in local or regional institutions rather than universities in the United States or Europe.

Turning from basic research to applied technology, anthropologist Grace Goodell (of the Program on Social Change and Development at the Johns Hopkins University School of Advanced International

Studies) challenged the research community to take the initiative in bridging the gap between theory and practice. Goodell wrote that scientists are abrogating their responsibilities by formulating IPM-related technologies that are too complex and impractical, relying on extensionists and even the farmers themselves to wade through this complexity and come up with practical methods. When they cannot, she claims, it is commonly regarded as the failure of the practitioners, whereas it should be seen as a failure of the science itself. From this perspective, it is up to the researchers to distill their results and simplify their recommendations to the point where they can be adopted in the field by the average farmer.

It is hard to argue with the principle. Clearly the sciences of entomology and ecology cannot be translated into the technology of pest management without due attention to the intellectual, physical, economic, and practical capabilities of the client group. Thus, for example, instructions for pesticide dosages can be given in terms of amount of the product to be used per sprayer tankful rather than per hectare, or economic thresholds can be expressed in terms of easily recognized and registered signs such as number of uninfested plants per infested plant.

The critical problem is to define the limits of researchers' accommodation to farmers and vice versa. For example, despite Goodell's proposals to the contrary, it seems inescapable under IPM that farmers must sometimes be able to recognize five or six (or even more) insects in key life stages in order to diagnose pest problems and respond accordingly. The same holds for significant weeds, and even for diseases that may often be confused with other problems such as nutrient deficiencies or cold. Similarly, even if they are not called upon to perform sophisticated economic cost/benefit calculations, farmers or their scouts must master basic random sampling techniques and analysis to assess levels of potential crop damage. These aspects are basic to IPM and cannot be eliminated or simplified to the point where no learning or accommodation on the part of the farmer is necessary. They can be more standardized, but only at the cost of the flexibility and responsiveness to the local situation, which is also an important aspect of IPM (i.e., an economic threshold of one borer per 10 heads may be appropriate under one set of circumstances, but not another).

As usual, the issue comes down to a question of degrees rather than absolute. Despite Goodell's call for increased emphasis and spending on extension at the expense of research, the critical need for continued, on-site research is clear. The experience of the recent Sahelian regional IPM project, funded by the U.S. Agency for International Development, provides a good example. In this project, insufficient time and support were given to research, with the result that, ultimately, the second-phase extension effort failed through a lack of appropriate new technologies to extend.

The need for effective, two-way communication among researchers, extensionists, and farmers has by now become so widely accepted as to become almost a platitude. Certainly a reorientation on the part of researchers is important, so that they take greater account of farmers' real needs, capabilities, constraints, and traditions. While an emphasis on research conducted by local scientists may seem to promote such an emphasis naturally, this is not necessarily the case. Local researchers can be just as wrapped up in abstract science, and just as seemingly unaware or even indifferent to real problems, as their expatriate counterparts. The solution is for research priorities to be developed from the bottom up, that is from the ultimate user to the researchers, and researchers must become very familiar with the entire farming system and not only the particular crop/pest interaction they are addressing. This should enable scientists to avoid promoting unworkable recommendations, such as early planting when the ground is too wet or too dry to plough, or burning crop residues needed for building materials or fodder. Researchers should also routinely investigate traditional methods of farming and pest control, to determine which of these might be important and valuable tools upon which to build.

The development assistance community is in fact responding to this common sense approach. It is becoming far more common for sociologists and anthropologists to be involved in the design of agricultural development programs, and for projects to emphasize two-way interaction and exchange of information among researchers, extensionists, and farmers. There is also a greater emphasis on farmers' and extensionists' involvement in the research process itself, through the use of on-farm trials and demonstrations. Lately, development agencies have even begun to recognize that the majority of African farmers are women, and that technologies and extension programs must be geared accordingly.

To enhance the effectiveness of IPM in developing countries, a balance must be struck between accepting the inherent complexity of ecological interactions, and meeting the need to provide relatively simple solutions to practical problems. This requires focusing on fundamentals rather than refinements. In most cases the critical first step in introducing an IPM approach is likely to be establishing an understanding of the role and significance of natural enemies in the agroecosystem, and identifying the measures required to protect them. For example, this has been the approach of the very successful Food and Agriculture Organisation's Integrated Rice Pest Control project in Southeast Asia, where the project team has put its greatest emphasis on training farmers to recognize beneficial insects and evaluate the importance of natural enemies in controlling rice pests in their fields. The critical role of spiders in controlling insect pests in rice paddies has become particularly clear through this project.

Protecting biological control agents is a fundamental tenet of IPM that applies in any situation, and the need for education in this area is great as many farmers do not have a real appreciation for their importance. This should therefore be a major focus of the state extension services. For maximum effect, education should be directed particularly at growers of crops that are currently the principle users of pesticides. Worldwide, this makes cotton a top priority. In the moist tropics, it includes rice, coffee, and cocoa, many fruits and vegetables, and, to an increasing degree, domestic commercial crops such as maize. Representatives of pesticide manufacturers are increasingly pervasive in developing countries, promoting their products, so extensionists must be there as well, helping farmers to use those products rationally and with minimum risk to themselves and to the environment, particularly to natural enemies of pest species. While many manufacturers do make an effort to provide some user training on pesticide safety, the one thing they cannot be expected to do is to help farmers use as little of their product as possible.

SPREADING THE IPM CONCEPT (REACHING THE NONECOLOGIST)

Nontechnical issues in implementation of IPM, such as distribution and control of land or economic incentives will ultimately determine whether IPM becomes a major force in tropical agriculture or is delegated to being a topic of interest primarily to academicians and research scientists. Therefore, while scientists are often more comfortable restricting themselves to technical matters, we also have an essential role to play in influencing the critical economic, and political decisions. Despite definite advances in the understanding and appreciation of the environmental bases of development among some developing-country officials, ecology and environmental science are still widely viewed as a luxury for the rich rather than as an essential issue for the poor, who are most directly dependent upon the natural environment. It is the responsibility of ecologists and entomologists to convince policy- and decision-makers of what we already know: that applying ecological principles to pest management is the only road to long-term gains in agricultural production, and that failure to do so has led, and will continue to lead, to disaster.

To accomplish this critical task of persuasion, ecologists must take the initiative to make their knowledge accessible to nonspecialists, by focusing on problems widely recognized as significant to human welfare, by publishing outside the technical literature and by presenting research problems and results in practical (often economic) terms. Only in this way, and by working closely with colleagues in other disciplines such as economics and sociology and in development agencies and governments, can we hope to see an increase in application of ecology to the problems and processes of economic development.

References Cited and Sources of Further Information

Allsopp, R., D. Hall, and T. Jones, 1985. Fatal attraction for the tsetse fly. New Scientist 18(1481):40–43.

Altieri, M.A., 1985. Developing pest management strategies for small farmers based on traditional knowledge. Bull. Inst. Dev. Anthropol. 3:13–18.

Chaboussou, F., 1986. How pesticides increase pests. Ecologist 16(1):9–35.

International Center for Insect Physiology and Ecology (ICIPE), 1984. Twelfth Annual Report. Nairobi, Kenya.

Metcalf, R.L., 1984. Trends in the use of chemical pesticides. *In:* Judicious and efficient use of insecticides on rice. Los Banos, Laguna, Philippines: International Rice Research Institute.

Metcalf, R.L., and A. Kelman, 1981. Integrated pest management in China. Environment 23(4):6–13.

Putnam, A.R., and W.B. Duke, 1978. Allelopathy in agro-ecosystems. Ann. Rev. Phytopathol. 16:431–51.

Repetto, R., 1985. Paying the price: pesticide subsidies in developing countries, Research Report No. 2. Washington, DC: World Resources Institute.

Risch, S.J., D. Andow, and M. Altieri, 1983. Agroecosystem diversity and pest control: data, tentative conclusions, and new research directions. Environmental Entomol. 12:625–629.

Sumner, D.R., B. Doupnik, and M.G. Boosalis, 1981. Effects of reduced tillage and multiple cropping on plant diseases. Ann. Rev. Phytopathol. 19:167–187.

Wellman, F.L., 1968. More diseases on crops in the tropics than in the temperate zone. Ceiba 14:17–28.

Chapter 5

DAMS AND DEVELOPMENT IN THE TROPICS: THE ROLE OF APPLIED ECOLOGY

PETER D. VAUX AND CHARLES R. GOLDMAN

The 1969 National Environmental Policy Act (NEPA) in the United States and the 1972 Stockholm Conference on the Human Environment emphasized the environmental consequences of economic development and our increasing awareness of it. Since the mid-1970s, for example, major international development institutions have required that some attention be paid to the environmental and social consequences of the projects they fund. This situation arose, in part, as a consequence of the U.S. Agency for International Development (AID) being forced to comply with NEPA (O'Riordan, 1981). Developing countries initially expressed a number of reservations about this trend, mostly related to the potentially negative economic impacts they thought might result. In the area of water resources development, resistance to involving environmental issues was still evident at the 1977 U.N. Water Conference in Mar del Plata. Nevertheless, during

101

the past decade, developing countries have slowly expanded their environmental interest.

Dam-building is a major component of many water-resources development projects. Almost 30 years have elapsed since the closure of the first large dam in the tropics (Kariba), and 20 years since the first international conference on man-made lakes focused on some of the ecological side effects of these projects. In spite of this passage of time and the post-Stockholm trends in environmental awareness, the environmental impacts of tropical impoundments, and the inadequate degree to which ecology is being incorporated into reservoir planning and development are a continuing concern.

Every reservoir project creates a dual need: to minimize negative environmental, social, and economic impacts; and to maximize positive contributions that can frequently be made both at local and national levels. To achieve these two objectives, applied ecology must focus on four components of reservoir development: pre-impoundment environmental impact assessment, ecosystem conservation and protection, post-impoundment auditing of the assessment process, and integrated environmental management of the reservoir and its basin. Furthermore, these four components must be carefully adapted to the specific demands and inherent constraints of the project in question.

In this paper, we examine the involvement of ecology in tropical hydroelectric projects primarily from the planning and management perspective; the environmental impacts, per se, of dam construction are here of secondary concern. Although our focus is on hydroelectric impoundments, much of the discussion relates to the broader field of water-resources development in the tropics.

HYDROPOWER DEVELOPMENT IN THE TROPICS

Major impoundments first started to appear in the tropics in the late 1950s, and, although the construction of large dams (as classified by Mermel, 1986) worldwide probably peaked in the early 1970s, substantial numbers are currently under construction or at the planning stage in the tropics. Approximately 35 percent of all large dams already installed or planned for commissioning between 1980 and the end of the present century are located within the tropics. Between 1980 and 1990, hydro capacity in the Third World—which includes most of the tropical countries—is projected to double, increasing from 100,000 megawatts to 201,000 megawatts when China is excluded. For a number of these countries, particularly in Latin America, hydropower already contributes a substantial share to their total generating capacity (Table 1). Nevertheless, considerable potential remains to be developed. By 1980, for example, Latin America and Africa had harnessed 8 percent and 5 percent, respectively, of their hydropower potential. Even in Brazil, where dams are already the major source of electrical power, installed capacity at the beginning of the 1980s represented only 12 percent of the country's 213,000 megawatt hydro potential. In contrast to the

TABLE 1

ELECTRICITY-GENERATING CAPACITY IN SELECTED DEVELOPING COUNTRIES OF THE TROPICS, 1982

Country	Total Capacity	Percent of Capacity Provided by:			
	(*megawatts*)	*Fossil* *Fuel*	Hydro	Geothermal	Nuclear
Brazil	38,904	15	85	0	0
Argentina	13,460	48	49	0	3
Bolivia	508	44	56	0	0
Colombia	4,660	36	64	0	0
Costa Rica	657	30	70	0	0
Nicaragua	400	65	26	9	0
Mexico	21,574	68	31	1	0
Zimbabwe	1,192	41	59	0	0
Kenya	574	33	62	5	0
Zaire	1,716	3	97	0	0
Senegal	165	100	0	0	0
Niger	100	100	0	0	0
India	38,808	64	34	0	2
Bangladesh	990	92	8	0	0
Thailand	4,694	71	29	0	0
Indonesia	3,513	83	16	1	0
Philippines	5,054	64	25	11	0
China	72,360	68	32	0	0

SOURCE: World Bank (1985): Power/energy data sheets for 104 developing countries (Flavin, 1987).

situation in Latin America and Africa, North America and Europe had already developed 59 percent and 36 percent, respectively, of their potential by 1980.

Although development of hydropower in the tropics will continue, there has been a gradual shift away from the very large impoundment projects characteristic of the past three decades, toward smaller-scale hydroelectric schemes. Not only are the environmental impacts of such projects frequently less severe than those arising from large impoundments, but also the construction cost per unit generating-capacity is often significantly lower. The potential for small-scale hydropower development in many parts of the tropics is considerable. For example, it has been estimated that in Peru and Costa Rica, this potential represents 364 percent and 415 percent of the total 1982 installed, generating capacity (from all sources); equivalent values for India and the Philippines are 28 percent and 83 percent (Flavin, 1987).

Thus the need for ensuring a more effective integration of applied ecology into tropical impoundment projects stems not only from past experience in this area, but also from the future role that hydropower will play in the economic development of many parts of the tropics.

ENVIRONMENTAL IMPACTS OF TROPICAL DAMS

A number of reviews have addressed the environmental consequences of dam construction (e.g., Goldman, 1976; Goldsmith and Hildyard, 1984; Petts, 1984; Goodland 1985, 1986, 1989). By way of summary, however, we have grouped major ecological/socioeconomic impacts into two broad categories (Table 2).

At level I, are the more obvious, direct effects such as the necessity for population relocation, the flooding of agricultural land, and the transition from a riverine to a lacustrine fishery. Resettlement of people living in the basin of a future reservoir is a complex task involving a broad spectrum of cultural, socioeconomic, and ecological considerations. If sufficient importance is given to the full range of issues, then resettlement is frequently also very expensive and one that may significantly alter the overall economic feasibility of the project if fully involved in the cost/benefit evaluation. The poor results of many resettlement schemes stem both from the complexities of planning and, especially, implementation of resettlement policies, as well as planners' frequent unwilling-ness to adequately recognize the economic consequences of enforced relocation (Goldsmith and Hildyard, 1984). Poorly developed political organization within the resettlement groups can compound these prob-

---

TABLE 2

EXAMPLES OF MAJOR POTENTIAL ENVIRONMENTAL IMPACTS OF TROPICAL IMPOUNDMENTS

Level I Impacts (Direct)

- Displacement of human population
- Changes in communications: road-building, lake transportation, cutting off traditional communication routes
- Loss of agricultural land
- Loss of archaeological sites
- Loss of wildlife habitat
- Increased incidence of some water-related diseases
- Water quality concerns: relevance to engineering structures, public health and reservoir fishery
- Aquatic macrophyte infestations
- Sedimentation in reservoir
- Change from lotic to lentic fishery
- Decrease in downstream fish production
- Various impacts on coastal ecosystems

Level II Impacts (Indirect)

- Immigration into area, and effect on watershed land-use practices, sedimenta-tion rates, etc.
- Increased integration of basin population into national economy through opening of region following reservoir development
- Influence on local economy of boom phase of fishery; possible increase in wealth inequalities
- Impacts on fish-marketing patterns resulting from improved communica-tions, change in balance of subsistence/commercial fishing
- Conflicts between local and immigrant groups
- Indirect effects of increased disease incidence, e.g., reduced productivity

NOTE: Level I impacts are a direct consequence of impoundment, whereas those in Level II are a more indirect set of impacts.

lems by hindering people from voicing their opinions within the bureau-cratic institutions in charge of the project. The proposed Purari River Project in Papua New Guinea is one case in which local people organized into a group (the Purari Action Committee) to protest the negative impacts that river development would have on their resource base and on the cultural integrity of their society. In general, however, although the resettled population is most directly affected by a dam, it often has little political power to influence project development policies.

The loss of prime agricultural land is especially significant when dams are built on rivers that have broad floodplains. The proposed dam below Lokoja on the Niger River in Nigeria, for example, would flood some of the best agricultural land in the region. The annual inundation of the Niger floodplain plays a vital role in local small-scale seasonal farming by providing both water and silt-associated nutrients. This role is especially important in view of the high cost alternatives of artificial fertilizers and irrigation. Similarly, the loss of productive riverine fisheries is likely to be especially severe when rivers with broad floodplains are impounded. Although reservoir fish production often represents a significant benefit of a project, the *net* benefit of the future reservoir fishery will vary with the nature of the river being impounded.

A second set of impacts (Level II) includes those that tend to be more diffuse, arising indirectly from river impoundment (Table 2). One example is the increased pressure on land resources in a reservoir basin that results from the influx of people into the area as a response to improved communications or changes in the economic base of the region. An additional example is the medium- to long-term effects on various sectors of the local economy following the frequently substantial cash inputs derived from the boom phase of the reservoir fishery.

Many of the impact types in Table 2 tend to be common to many impoundments throughout the tropics. In some cases, however, major results can be highly specific to the particular project in question. One example concerns a proposal to dam the Parana River in Argentina, which would involve a massive diversion of water through the 1,500-square-kilometer Ibera swamp. Initial environmental assessment studies concluded that this diversion would raise water levels in the wetland sufficiently to uproot vast areas of loosely attached plants. These would then collect at discharge structures and cause severe problems in the operation of the system. A somewhat analogous problem has occurred at the Arenal reservoir in Costa Rica where sections of a flooded peat marsh have floated to the surface and then been driven by the wind toward the intake structures of the hydroelectric plant. The resulting blockage and damage to the intakes have shut down the generating plant at least once.

WHY FOCUS ON TROPICAL IMPOUNDMENTS?

While it is perhaps not wise to delineate decisively between tropical and temperate reservoir ecosystems in terms of their structure and function, tropical impoundments deserve special attention for a number of reasons, apart from their overall contribution to present and future dam-building activity. These include the following:

1. The legal and administrative infrastructure for incorporating environmental planning into engineering projects is frequently less well-defined in many of the developing countries of the tropics than it is in the more industrialized nations.

2. Financial, logistic, or scientific resources that can be used for developing and promoting this involvement of applied ecology in water-resources development projects are also frequently less readily available in many tropical countries. This makes it especially important for those resources that are available to be used as efficiently as possible.

3. The secondary benefits of tropical impoundments can be significant and, in view of the frequently low per capita income in surrounding areas, urgently need to be maximized. This is especially so given the fact that the primary beneficiaries of impoundment projects (consumers of hydropower, for example) are often those sectors of the community, usually urban, which are relatively distant from the reservoir. Many of the negative effects, on the other hand, fall largely on people living in the immediate lake basin.

Fisheries are a prime example of secondary benefits, since they potentially represent an important source of both protein and employment in areas where malnutrition and underemployment or unemployment are common. This contrasts with the usual situation in the more developed economies where freshwater fish is either a highly priced luxury or the object of a leisure-time sport industry. Several examples from tropical hydroelectric projects demonstrate that the value of the fish harvest has actually exceeded the net income from power generation. In Ghana, for example, the Volta Lake fishery was worth approximately twice the surplus income derived from the sale of electricity in 1978. In Thailand, income from fish harvested from the Ubolratana reservoir has exceeded that from power generation by 20 percent. Similarly, fish production from the Nyumba ya Mungu reservoir in Tanzania, during the first decade following its impoundment, had an accumulated value almost equal to the cost of dam construction itself. Although reservoir fish production eventually decreases from the peak levels characteristic of the immediate post-impoundment period, the high incomes generated within fishing communities during the boom phase of the fishery can enter and stimulate various sectors of the local economy. As a result, ramifications of the peak phase of the fishery can still be seen in a region many years after fish production has declined.

Not all tropical reservoirs, however, experience the fishing pressure that has characterized many of the larger African impoundments. At the El Cajon reservoir in Honduras, for example, the reservoir's steep sides combined with a poorly developed inland fishing culture to produce only a slow development of the fishery on this new impoundment. Nevertheless, various estimates of potential fish yield from the El Cajon reservoir suggest that the fishery, if fully developed, could contribute significantly to the protein intake of the surrounding population.

4. A major consideration in determining the economic feasibility of an impoundment project is the amount of sediment build-up in the live storage of the reservoir. For many existing impoundments and potential dam sites in the tropics, we have only relatively short-term data bases for river sediment loading. At the same time, changing patterns of human activity in many tropical countries—for example, intensification in the use of the traditional and widespread shifting agriculture (slash and burn)—can significantly increase erosion and sediment transport rates. These two factors mean that sustained and interdisciplinary attention must be given to the area of reservoir sedimentation. Since the economic life of the project may be at stake, reservoirs represent a potentially valuable focus for developing and implementing programs of improved watershed management. Many parts of the tropics urgently need such programs as the pressure on land resources increases.

5. When an impoundment floods large areas of tropical forest, reservoir water-quality problems—for example, hydrogen sulfide generation—are likely to be especially severe, and may significantly affect the engineering structures of the dam itself. Many dams being planned for the tropics are located in densely forested regions, and are thus of particular interest in terms of water quality. Although of special concern in these cases, hydrogen sulfide generation can also be a problem in areas where the inundated forest cover is considerably less dense. The El Cajon dam, for example, recently impounded a 94-square-kilometer reservoir in an area of central Honduras that had already been heavily logged in the past and that today supports a relatively sparse cover of pine and oak forest. Hydrogen sulfide accumulated in the deep, hypolimnetic regions of the new reservoir and eventually leaked into the power house atmosphere via the turbine-cooling water intake. This corroded various electrical components, such as relays, which had to be replaced.

HISTORY OF ENVIRONMENTAL WORK ON TROPICAL RESERVOIRS

In general, the greatest financial investment in environmental research has been associated with the larger reservoirs, since these tend to be both more noticeable and more expensive. These impoundments also generate environmental effects on a larger scale (larger in an absolute sense—the relationship may not always hold if the effects are somehow expressed on a per capita or per unit area basis).

Ecological investigations of the major African impoundments, for example, have been well documented (e.g., Entz, 1984; Kapetsky and Petr, 1984). They range from the socioeconomic and fisheries surveys of the pre-Kariba Zambezi River basin, to the relatively large Food and Agriculture Organisation/U.N. Development Programme programs for the Volta, Aswan, and Kainji impoundments, to the more restricted study at Cahora Bassa. Many of these research programs were started prior to the mid-1970s trend toward increased environmental awareness and they have done much to increase our understanding of tropical reservoir ecosystems. However, most of the work has been carried out *after* the initiation of dam construction; on the whole, project assessment and design have not integrated the environmental dimension in detail. Although the Volta project in Ghana represents a good example of early planning to study the nonengineering aspects of dam construction, even in this case much of the environmental work was performed in the context of a salvage operation after various problems had arisen. The Ubolratana reservoir in Thailand is another example of this problem of poor timing. Here, a wide-ranging environmental program has been in operation for about 20 years, analyzing the benefits and continuing problems associated with this project. These studies, however, began at the same time as construction of the dam itself, and thus could not contribute to the planning process.

Two more recent examples from Latin America further illustrate this coincidence between the start of dam construction, on the one hand, and environmental studies on the other. In Brazil, the recently completed Tucurui dam will have a final installed generating capacity of 8,000 megawatts and impound the largest reservoir ever created in a zone of tropical rain forest (2,160 square kilometers at maximum surface elevation). Tucurui is one component of a vast development project aimed at exploiting the mineral and forest resources of the eastern Amazon region. Despite the impressive scale of the Tucurui project and the little amount that is known of the ecology and socioeconomic status of the surrounding area, an environmental impact assessment for the project was not requested by the Brazilian power company, Electronorte, until one year *after* the initiation of engineering work. Once this study (Goodland, 1977) had been completed, the only environmental problems to subsequently receive attention were those that directly affected dam construction and operation.

Similarly, in-depth environmental studies for the 300-megawatt El Cajon dam in Honduras started at the same time as dam construction, even though a brief survey of the area had drawn attention to some of the environmental aspects of the project seven years before. Thus, ecological considerations had little chance of being integrated into project assessment and design. In the context of tropical reservoir development, however, the El Cajon project represents one of a minority of cases for which there have been detailed pre-impoundment investigations of water quality, aquatic ecology, and fisheries. More important, it has also provided a significant stimulus for the development and establishment of an indigenous research effort associated with water-resources projects in Honduras. This means that applied ecology will probably play a greater role in the future development of hydropower projects in this part of Central America.

Even in those cases where ecology is addressed relatively early on in a project, it does not necessarily follow that this work significantly influences either the planning of the reservoir or its subsequent management. For example, social and environmental studies were included at the feasibility stage in planning Tanzania's largest dam, Mtera. Following impoundment, however, the series of measures designed to reduce negative impacts and promote the positive contributions of the project were ignored. Similarly, multidisciplinary studies on the Kafue basin in Zambia began in the 1960s, but have had little influence on major planning decisions to date.

Although there is justified criticism of the timing of the research effort, at least some environmental work has often been possible for the larger tropical impoundments, principally because these projects attract national or international funding. In contrast to these major projects, however, the thousands of smaller reservoirs scattered throughout the tropics are presumably in a much less favorable position to attract either local or outside funding for environmental research. As a result, it seems likely that applied ecology is rarely involved at any stage in the development of these projects.

THE ROLE OF ENVIRONMENTAL IMPACT ASSESSMENT

In any reservoir project, the first step for applied ecology is an environmental impact assessment. Ideally, the environmental impact study is designed to identify and quantify both negative and positive effects, consider possible mitigation measures, and ensure that the economic consequences of these impacts are incorporated as fully as possible into the project's overall feasibility assessment.

Procedures for a formal environmental impact assessment were largely developed in the United States following the passage of the 1969 National Environmental Protection Act, and have since been adopted to varying degrees by a number of other countries, including several in the tropics. Environmental impact studies, in general, have been criticized in the past, both in terms of failing to adequately fulfill their objectives (Hecky et al., 1984; Rosenberg et al., 1981; Schindler 1976), and for not using available impact assessment methods to the fullest extent possible (Canter, 1983). Another criticism has been that studies frequently attempt to collect too much information of questionable importance to the actual process of impact assessment (Rosenberg et al., 1981; cf. Petr, 1978). However, perhaps the major weakness of impact studies is the poor attention that is usually given to post-project audits (Rosenberg et al., 1981). In the case of reservoir projects, post-audits are based on post-impoundment evaluations of the environmental consequences of dam construction. In addition to providing case-history descriptions, the major function of these audits is to evaluate the predictions formed during the pre-impoundment phase of the project. Such predictions should, in effect, be a series of hypotheses that can subsequently be tested by post-impoundment audits. These audits thus evaluate the overall success of the pre-impoundment environmental impact assessment. Few comprehensive pre- and post-impoundment comparative analyses have been undertaken in the case of tropical reservoirs, even though the general literature on the ecology and limnology of these systems is increasing. Even fewer studies have evaluated in detail the post-impoundment situation with reference to the set of predictions made at the pre-impoundment stage. The need for these post-impoundment, hypothesis-testing assessments is becoming even greater as the development of more sophisticated lake and reservoir modeling techniques increases the predictive capabilities of both ecologists and engineers.

A number of the developing countries of the tropics have found that the U.S.-based procedures of formal environmental impact assessment are not especially well-suited to the demands and constraints in their own regions. Insufficient funding, lack of appropriate data bases, and lack of public participation and administrative infrastructure are some of the major problems. Instead of a rigid impact assessment method, these situations call for a broader and more integrated approach tailored to the particular needs and constraints of the particular country. This approach is essentially assessment *and* management oriented; it is designed to collect information that will not only provide for impact assessment but will also assist in optimizing benefits (primary and secondary) expected from the project in question. This philosophy does not, of course, absolve ecologists from arguing as forcefully as possible for a modification in or even a cancellation of projects they consider will result in excessively negative impacts. It does, however, provide for a more constructive approach to situations in which a realistic appraisal would suggest that the project is likely to go ahead in some form regardless of ecological input, and the negative impacts are unlikely to be overwhelming, while at the same time the secondary benefits are potentially significant.

In most situations, the options of either no development or only 100 percent environmentally sound development are unrealistic (Carpenter and Dixon, 1985). Therefore, the implementation of wisest possible development strategies depends on recognizing trade-offs (Goodland, 1986) and including as many relevant factors as possible in the subsequent cost-benefit analysis. Care must be taken not to let concern about negative results, justified as it might be, obscure the potentially positive aspects of a project. This is important since full benefits are frequently only realized if they are adequately identified and subsequently developed. Negative impacts, on the other hand, usually occur whether or not they have been predicted, monitored, or worried about. Apart from anything else, excessive concentration on negative aspects may prove counterproductive over the long term, since it may lead to a degree of wariness on the part of the engineering and decision-making communities that in turn will jeopardize the effective integration of environmental considerations into project development.

CONSTRAINTS AND PROBLEMS IN APPLIED ECOLOGY

Efforts to incorporate applied ecology into tropical impoundment projects frequently confront a wide range of constraints. We have grouped these into the three major categories of institutional, methodological, and ecological constraints (Table 3). Most of these factors are, of course, not specific to dam projects or, indeed, to the tropics.

TABLE 3

CONSTRAINTS TO THE INTEGRATION OF APPLIED ECOLOGY WITH RESERVOIR DEVELOPMENT IN THE TROPICS

Institutional

- Lack of infrastructure for requiring environmental work at planning stage and/or for incorporating findings into project design and execution.
- Project inertia; difficulty in arresting development, or changing direction, of something that has already started, especially when national prestige is involved.
- Frequently there is a pre-supposed consensus about the need for the project; the do-nothing option is not considered.
- Short-term gains frequently mask long-term consequences: governmental and/or public perception of and emphasis on environmental dimension is often less strong than it is on short-term economic advancement.
- Economic base of reservoirs frequently emphasizes primary factor(s), such as power production, irrigation potential, etc., rather than including all components that will have an associated cost—e.g., resettlement, food subsidies, health costs, etc.
- Environmental problems are not considered until serious economic impacts arise.
- Inadequate funding.
- Insufficient personnel, especially those with experience in impact studies.

Methodological

• Problem of assigning economic values, or some other comparative indicators, to social and environmental factors.

• Assessment methodology: choosing the most appropriate system for the situation in question.

• Studies may be too wide-ranging for the resources at hand: leads to either collection of data of no particular use for impact study or insufficient data on topics that are of immediate concern, i.e., resources spread too thinly.

• Recommendations may be too wide-ranging and, in effect, impractical to carry out (although quite possibly desirable); may be better to keep scale smaller and objectives realistic.

Ecological

• Difficulty of making predictions for ecological systems: interrelationships between components of a system often not well understood.

• Inadequacy of pre-impoundment data base, in terms of either time-scale or parameters considered.

• Paucity of pre-/post-impoundment comparative case history studies.

• Insufficient information about, or to take into account, potential future events that will influence prediction set; for example, influence of higher rates of immigration into an area and effects on erosion rates.

The poor timing so often characteristic of environmental impact assessment studies is often partly a result of institutional considerations. Two factors that impoundment projects frequently develop are especially important here: the absence of a well-developed infrastructure that ensures that impact assessments are carried out—this infrastructure may have a strictly legal basis, or it may depend more on convention; and the impetus for development and completion. The latter factor, in particular, has been common in the past and continues to be so today. The glowing economic arguments made for many of the large dams, together with consequent national prestige and their identification as major instruments of development (effectively equating construction with development), tend to make it much more difficult to objectively include side-issues such as environmental evaluations in the process of economic feasibility assessment. Even when pre-impoundment assessments are carried out, this same project momentum can be responsible for politicians' and planners' refusal to reconsider the full implications (economic, social, and environmental) of proceeding with construction.

Methodological factors (Table 3) also contribute to the inadequate incorporation of environmental assessments into reservoir planning and development; frequently many uncertainties and difficulties are associated with the actual process of predicting and evaluating project impacts. Chief among these is the question of monetarizing, or assigning realistic and comparative values to each of the predicted costs and benefits. This is especially so for some of the more "diffuse" impacts (Table 2, Level II), such as indirect costs associated with the increased incidence of disease following impoundment. The spread of malaria and schistosomiasis (biharzia), for example, not only has direct costs such as those for vector control and drug treatments; it also has the indirect costs of decreased human productivity, or work potential, and increased vulnerability of people to other diseases.

The loss of forest resources by flooding is another example of the problem of value assignment. Estimating the commercial or firewood value (on the basis of sustainable exploitation) of an area of forest is a relatively straightforward process, at least in theory. This same forest, however, may also provide significant habitat for rare or endangered species, or may be deemed important because this type of ecosystem is becoming scarce elsewhere. Alternatively, it may simply have great aesthetic value. In these situations, monetarization and full value assessment are considerably more difficult. Sometimes, these difficult-to-quantify costs are intuitively high enough, or of enough public concern, to argue forcefully to influence the planning process on its own strength. A notable example here is the Silent Valley project in southern India, where the proposal was to dam the Kunthipuzha River for power and irrigation. The dam would have flooded 540 hectares of tropical rain forest within the already constituted 8,952-hectare Silent Valley Reserve Forest. Although raising many other objections to the project, environmentalist groups focused on this forest loss and argued successfully against completion of the project, even though 3 million U.S. dollars had already been spent on it. The case of Silent Valley is almost unique in the tropics since the project was halted at a late stage in development, entirely for environmental reasons. In many other situations, single components of the issue are not as clear-cut or cannot attract such a degree of public participation. Here, decision-making must rely entirely on a rigorous and inclusive cost/benefit analysis.

The positive impacts of an impoundment project can also be difficult to quantify realistically. Reservoir fisheries provide a good example. Although methodology is available to make pre-impoundment estimates of post-impoundment fish production (the morpho-edaphic index in particular), the actual fish harvest from a reservoir depends not just on biological fish production but also on a spectrum of socioeconomic factors. These include the development of marketing patterns and the improvement of fishing technology to allow effective usage of the open waters of the new lake. In Kainji Lake, Nigeria, for example, fish harvest is considerably lower than the maximum sustainable yield, because the fishermen lack appropriate gear and boats to take fish that inhabit very deep water. Such factors as these make it much more difficult to assess fish *yield* than fish *production*, especially over the medium to long-term range.

In addition to the problems of value assignment, some problems are essentially ecological in origin (Table 3). Making good predictions for ecological systems is usually inherently difficult since many of the component interrelationships are often poorly understood. The frequent inadequacy of pre-impoundment data bases and the paucity of detailed case histories of post-impoundment audits also complicate impact prediction. The fact that it is often not possible to make precise ecological predictions in as clear-cut terms as those generated in economic and engineering feasibility studies works against an effective incorporation of the environmental dimension into project appraisal and planning. Some of the components of conventional economic analyses for tropical impoundment projects—for example, forecasts of power demand, potential generating capacity, and the possibilities for developing irrigated agriculture—have been highly unrealistic on more than a few occasions. However, the appearance of diffuseness that often characterizes ecological predictions means that these are often less likely to be taken seriously by planners, politicians, and engineers who are accustomed to more cut-and-dried arguments about project feasibility.

Many of the constraints to pre-impoundment environmental impact assessments also influence the post-impoundment applications of applied ecology (post-audits and reservoir basin management). The availability of adequate funding is of obvious interest in this respect. Dam construction has provided a major stimulus for a significant proportion of the aquatic ecological research undertaken in the tropics over the past 30 years, even though, as we have already seen, this environmental work typically begins at a late stage in project development. Completion of the dam, however, may actually signal the end to one of the major forcing factors for ecological work—that is, environmental assessment as a loan requirement of funding institutions. In other words, once the phase of high-activity, high-investment dam construction has ended, it may often be much harder to attract sufficient interest and funding from national or international organizations to ensure that necessary management-oriented research is maintained on the new reservoir.

INTEGRATING APPLIED ECOLOGY INTO DAM PROJECTS: PROSPECTS AND NEEDS

One of the consequences of the constraints discussed above has been that many impact assessment studies have not been true assessments at all. Instead, they have frequently taken the form of descriptive accounts of the pre-impoundment environment including, perhaps, a list of predicted consequences of project implementation. While this is obviously preferable to ignoring ecology altogether, it definitely falls short of providing the sort of information that will lead to a full integration of environmental considerations into project design, appraisal, and management.

These constraints emphasize the need to plan assessment procedures in relation to the specific context in which they are being used so that they actually serve some function. Pre-impoundment environmental investigations will fall into one of two broad categories that, in effect, represent opposite ends of a relatively continuous spectrum. The first group includes true impact assessment studies, which analyze and subsequently rank a number of possible development options according to their desirability in terms of economic and environmental considerations. The second group comprises the more descriptive studies that provide baseline data; these studies may or may not make a series of predictions about the impacts of reservoir formation, but if they do, the predictions are not integrated into a comparative cost/benefit analysis of the project. Within either category, of course, the scale and scope of a particular study can vary considerably. Investigations in both categories can theoretically be carried out at any stage during project development and implementation, but true impact assessment studies will be of direct use only if they are done during the planning stages. The more descriptive studies at the other end of the spectrum will generate baseline data even if they occur relatively late during project development, for example, after dam construction has begun.

A difficulty arises, of course, in deciding what is feasible in a particular situation. In-depth anticipatory environmental work will more likely occur in regions that have some tradition of requiring environmental impact assessments (whether or not this has an enforceable legal basis), or in projects large enough to require substantial levels of international funding. Public participation during the planning stages can also act as a catalyst for environmental assessment. Although there is not, in general, a very notable history of public involvement in the planning of tropical reservoir projects, the proposed Tembeling dam in Malaysia and the Silent Valley project in India are two examples of projects that were substantially influenced by nongovernmental, environmentalist pressure groups.

The possibilities for adequately addressing the environmental costs of a project are also likely to be influenced by the perceived economic or political needs for that project and the general economic situation within the country at the time. It will be much easier to achieve a full environmental analysis in situations without an overbearing economic or political pressure to complete the dam as quickly as possible, or at times when the national economy is robust enough to allow resources to be directed toward the supposedly fringe aspects of the project.

Although it is clear that many gaps remain in ecologists' understanding of the functioning of reservoir ecosystems, it is our opinion that, even with the existing base of experience, there is considerable room for improving the integration of applied ecology into tropical reservoir development. This process will rely to a large extent on four primary factors:

• *An improvement in pre-impoundment assessment method, particularly in assigning economic, or some other comparative value judgment, to as many factors as possible.* For example, in assessing the need for clearing forest from the reservoir basin before impoundment, the cost of this operation (in regions of dense forest, it may be very high) must be weighed against the likely costs of poor water quality (low dissolved oxygen, high hydrogen sulfides), which would affect engineering structures in the dam as well as public health.

Similarly, economic analyses must be carried out to estimate whether adjusted patterns of dam management could result in a net economic gain for the country (large dams, especially, need to be viewed as national resources, not just as the domain of power companies). Two examples here are: ensuring the regularity of the annual drawdown and filling cycle of the reservoir, which will in turn permit the development of drawdown agriculture; and releasing water from the reservoir to create artificial floods on the downstream floodplain, and promoting the survival of the fisheries of this region. Both drawdown agriculture and floodplain fisheries can be highly productive in economic terms, as well as for their food supply and employment values. Because of the lagtime between rainfall and reservoir filling in the larger reservoirs, crops produced by drawdown agriculture often reach markets at a time when those produced by nondrawdown, nonirrigated agriculture are scarce, and prices consequently high.

In the case of hydroelectric projects, however, developing both drawdown agriculture and the downstream fishery have economic costs in terms of lost power production, since both can entail a reservoir management policy that is not consistent with maximizing the impoundment's storage function. For example, unexpectedly high floods entering the reservoir during the early part of the wet season must not be allowed to raise the water level to a stage where drawdown crops will be flooded. This extra water thus has to be released from the reservoir without generating power. Similarly, mimicking the pre-impoundment flood regime in the downstream section of the river will entail releasing additional quantities of water at a time (beginning of the wet season) when all available water over and above power generation requirements would normally be stored. In many tropical reservoirs, the costs of lost power production may be more than compensated by the benefits of increased food production. Such integrated cost/benefit analyses are urgently needed for reservoir projects in the tropics—preferably at the planning stage, but, if necessary, after impoundment has occurred.

• *The effective execution and dissemination of a greater number of post-audit studies.* This will increase our case history knowledge of both impoundment ecology and the assessment process.

• *Greater emphasis on the longer-term consequences of dam construction in the tropics.* In Canada, more than 85 percent of the environmental budget has been spent on the preparation of environmental impact statements, leaving little additional funding for follow-up studies of the completed projects. A similar pattern in which a majority of the investment goes to near-impoundment (often too late to be truly pre-impoundment) environmental work has undoubtedly occurred, and continues to occur, for many projects in the tropics.

• *A careful orientation by ecologists to the issues of major importance for the project in question.* Research efforts must also be integrated so that, for example, reservoir limnology is not studied outside the context of what is going on in the watershed. The linkage between reservoir sedimentation and changing patterns of human activity in the lake basin is a prime example here. Much environmental management in developing countries has been characterized by an uncoordinated and unsystematic approach. Only by improving this situation can ecologists in many of these regions (where resources for environmental research are generally very scarce) develop a position where they can significantly influence planners and decision-makers. Even small research units can have a substantial effect at national and local levels if investigations are wisely oriented.

In order to become a reality, environmentally wise development in the tropics will inevitably require a sufficiently receptive infrastructure —with its political, legal, and bureaucratic components. To a large extent, however, it is also the responsibility of ecologists to promote this type of development by ensuring that their work addresses real questions in a convincing manner, and also that their conclusions are effectively communicated to planners and decision-makers.

Dam building in the tropics will continue to have a significant effect on the environment and socioeconomics of many tropical countries over the next few decades. Although reservoir projects have been a stimulus for the development of aquatic ecological research in various parts of the tropics, applied ecology has rarely played a significant part in project design and feasibility appraisal. Environmental impact assessment techniques must be modified to take into account the particular conditions and constraints in most tropical countries. In addition, more critical post-impoundment case history reviews are needed both of the consequences of dam construction and of any pre-impoundment assessment work carried out.

Although many gaps remain in our understanding of the structure and functioning of tropical reservoir ecosystems, much can be done to increase the effectiveness of the role applied ecology plays in reservoir development and management policies. One of the greatest needs is to view reservoir projects (especially large ones) within a broader, multiple-use context, thereby improving the degree to which the spectrum of ecological impacts is brought into a project's overall cost/ benefit feasibility analysis.

The future of applied ecology within reservoir projects, in particular, and economic development in the tropics, in general, will also depend to a significant degree on how effectively ecologists communicate their findings (and concerns) to planners, engineers, bureaucrats, and the public. Education must play an important part in further developing the role of applied ecology.

References Cited and Further Sources of Information

Allen, R.N., 1972: The Anchicaya hydroelectric project in Colombia: design and sedimentation problems. *In:* M. Taghi Farvar and J.P. Milton (eds.), The careless technology. Garden City, NY: Doubleday, Natural History Press, pp. 318–342.

Anonymous, 1983: Environmental impact assessment: problems and propects. Water power and dam construction 35(7):31–36.

Anonymous, 1983: The storm over Silent Valley. Water power and dam construction 35(7):14–17.

Biswas, A.K., 1986. Environment and law: a perspective from developing countries. Environmental conservation 13:61–64.

Canter, L.W., 1983: Impact studies for dams and reservoirs. Water power and dam construction 35(7):18–23.

Carpenter, R.A., 1981: Balancing economic and environmental objectives: the question is still how? Environmental impact assessment review 2: 175–188.

Carpenter, R.A., and J.A. Dixon, 1985. Ecology meets economics: a guide to sustainable development. Environment 27:7–11, 27–32.

Dasmann, R.F., J.P. Milton, and P.H. Freeman, 1979: Ecological principles for economic development. New York: Wiley, 252 pp.

Entz, B.A.G., 1984: A synthesis and evaluation of activities of FAO/UNDP projects of five African man-made lakes: Kainji, Kariba, Kossou, Nasser-Nubia and Volta, Fish. Circ. 774. Food and Agriculture Organisation, 34 pp.

Fernando, C., 1980: Tropical reservoir fisheries: a preliminary synthesis. Tropical ecology and development 1980:883–892.

Flavin, C., 1987: Electricity in the developing world. Environment 29:12–15, 39–45.

Freeman, P., 1974. The environmental impact of a large tropical reservoir: based on a case study of Lake Volta, Ghana. Washington, DC: Smithsonian Institution, 86 pp.

Goldman, C.R., 1976. Ecological aspects of water impoundment in the tropics. Rev. Biol. Trop. 24 (suppl.1):87–112.

Goldman, C.R., and R.W. Hoffman, 1977. Environmental aspects of the Purari River scheme. *In:* J.H. Winslow (ed.), The Melanesian environment. Canberra, Australia: Australian University Press, pp. 325–341

Goldman, C.R., and P.D. Vaux, 1984: El Cajon Hydroelectric Project, Limnology and Fisheries Program, final report, vol. 1: Limnology. Switzerland / E.N.E.E., Honduras: Motor Columbus Consulting Engineers, 575 pp.

Goldsmith, E., and N. Hildyard, 1984. The social and environmental effects of large dams. Vol. 1: Overview. Cornwall: The Wadebridge Ecological Centre, 346 pp.

Goldsmith, E., and N. Hildyard, 1986. The social and environmental effects of large dams. Vol. 2: Case studies. Cornwall: The Wadebridge Ecological Center, 331 pp.

Goodland, R.J. 1977: Environmental impact assessment: Tucurui hydroelectric project, R. Tocantins. ELECTRONORTE, Brazil.

Goodland, R.J., 1985. Environmental aspects of hydroelectric power and water storage projects. *In:* Environmental impact assessment of water resources projects, U.P. Roorkee, India, vol. 3. UNESCO/U.N. Environment Programme, pp. 1–30.

Goodland, R.J., 1986. Hydro and the environment: evaluating the tradeoffs. Water Power and Dam Construction 38:25–28, 33.

Goodland, R.J., 1989. The World Bank's new policy on the environmental aspects of dams and reservoirs. Washington DC: World Bank, 31 pp.

Hecky, R.E., R.W. Newbury, R.A. Bodaly, D. Patalas, and D.M. Rosenberg, 1984. Environmental impact prediction and assessment: the Southern Indian Lake experience. Can. J. Fish. Aquat. Sci. 41:720–732.

Holling, C.S., 1978. Adaptive environmental assessment and management. New York: Wiley, 377 pp.

Kapetsky, J.M., in press. Management of fisheries in large African reservoirs—an overview. *In:* Proceedings of the National Symposium on Managing Reservoir Fishery Resources. American Fisheries Society.

Kapetsky, J.M., and T. Petr, 1984: Status of African reservoir fisheries. CIFA Tech. Pap. 10.

Lowe-McConnell, R.H. (ed.), 1966. Man-made lakes. New York: Academic Press. 218 pp.

Mermel, T.W., 1986. Major dams of the world: 1986. Water power and dam construction 38:33–39.

O'Riordan, T., 1981. Problems encountered when linking environmental management to development aid. The Environmentalist 1:15–24.

Petr, T., 1975: On some factors associated with the initial high fish catches in new African man-made lakes. Arch. Hydrobiol. 75:32–49.

Petr, T., 1978: Tropical man-made lakes: their ecological impact. Arch. Hydrobiol. 81:368–385.

Petts, G.E., 1984: Impounded rivers. New York: Wiley, 326 pp.

Rachagan, S.S., 1983: Malaysia abandons Tembeling. Water power and dam construction 35(7):43–47.

Rosenberg, D.M., et al., 1981. Recent trends in environmental impact assessment. Can. J. Fish. Aquat. Sci. 38:591–624.

Schindler, D.W., 1976: The impact statement boondoggle. Science 192:509.

Vargas, E., and P.D. Vaux, in press. Limnological and ecological considerations for hydropower development in Honduras. Lake and reservoir management.

Vaux, P.D., and C.R. Goldman. 1985: El Cajon hydroelectric project, limnology and fisheries program, final report, vol. 2: Fisheries. Switzerland/ E.N.E.E., Honduras: Motor Columbus Consulting Engineers, 286 pp.

White, G.F., 1972. Organizing scientific investigations to deal with environmental impacts. *In:* M. Taghi Farvar and J.P. Milton (eds.), The careless technology. Garden City, NY: Doubleday, Natural History Press, pp. 914–926.

Chapter 6

TEACHING APPLIED ECOLOGY TO NATIONALS OF DEVELOPING COUNTRIES

Asanayagam Rudran, Chris M. Wemmer, and Mewa Singh

In 1980 the *World Conservation Strategy* identified the lack of trained personnel as a major limitation for the implementation of environmental conservation measures in developing countries. In Indonesia for instance, there were only 400 trained foresters, or just one forester per 3,000 square kilometers of forest. In order to overcome this widespread problem the *World Conservation Strategy* suggested that programs for training and research in environmental conservation be initiated in developing countries. In 1985, the U.S. strategy on the conservation of biological diversity reiterated the need for training natural-resource management personnel and stressed the importance of strengthening the technical capabilities of developing-country institutions that deal with environmental conservation. In 1988, the World Bank's wildlands policy again highlighted the importance of personnel training for the proper management of wildlife habitats. Thus during the present decade highly influential agencies dealing with international economic development and environmental conservation alike have spoken out in support of training in natural resource management. Such support

125

reflects the growing realization that natural resource management must be integrated with the process of economic growth in developing countries. Hopefully, it also signals the end of an era when conservationists, developers, and scientists of diverse disciplines were involved in acrimonious debate over the ethics, priorities, and future of mankind.

Now that support for training in natural resources management has been corralled, effective programs to provide such training should be established without delay. Present circumstances warrant this urgency. Over the last several decades rapid economic growth in developing countries was translated into the indiscriminate alteration of natural habitats. As a result, it is currently estimated that tropical forests the size of New York City disappear every four days and 1,000 of their species become extinct each year. At this rate of loss of habitats and species, all tropical forests would disappear forever in the next 80 years. These statistics, although estimates, give at least a general idea of the nature and extent of the problem of natural resources management in developing countries. Compounding this problem is the likelihood that the push toward rapid economic development will continue unabated for the next several years. Therefore, personnel training programs in natural resources management should be implemented immediately in developing countries.

But, implementing personnel training programs cannot alone bring about the integration of natural resources management and economic development. The reason is that applied ecology, which provides the scientific basis for the techniques of natural resources management, is a relatively unknown aspect of biology in developing countries. Some developing-country nationals may be familiar with the principles of ecology but they rarely realize the practical value of these principles. A few may be aware of the concept of an animal's home range but hardly anyone would understand its relationship to the creation of corridors (or protected areas) for the animal's movements. This should not be too surprising because, even in Western countries, applied ecology began to flourish only about two decades ago. In any case, the widespread ignorance of this science in developing countries makes the integration of natural resources management with economic development a very complex issue. Not only should we train natural-resources-management personnel, but we should also disseminate this important science extensively in developing countries so that it influences the thinking of a much wider audience, including decision-makers and politicians. Unless the latter task is accomplished, it will not be possible to integrate natural resources management with the process of development or avoid the negative environmental impacts of the current patterns of economic growth. Thus, the teaching of applied ecology to a diverse group of people in developing countries is vitally important. This poses numerous challenges. Clearly, a comprehensive plan of action must be worked out for teaching applied ecology to the people of developing countries.

THE CHALLENGES

In developing countries, so few people are knowledgeable or qualified to teach the subject of applied ecology that curricula for teaching this subject are virtually nonexistent at all levels of education. Even university biology graduates are unaware of the intimate relationship between man and natural resources and, consequently the intrinsic value of wildlife habitats. This creates a bias against seeking any form of employment related to wildlife habitats, which is exacerbated by the paltry wages paid by developing-country governments for work in these areas. These governments also pay biology teachers relatively poorly. Thus, the systems of education and government in developing countries have helped to create a misconception among people that wildlife habitats have no significant social value and the teaching of biology is relatively unimportant. This misconception leads to two problems. The first is that people who work in wildlife habitats are frequently those who become employed early in life because of financial necessity or lack of opportunity for a higher education; they tend to be relatively unsophisticated and lack the proper educational background for the conservation and management of wildlife habitats. The second problem is that a large proportion of those who graduate in biology move to jobs outside this discipline. After three to four years of expensive university education, only rarely will a graduate settle for a poorly paid teaching job. This in turn results in a dearth of qualified teachers and appropriate curricula in applied ecology and perpetuates a vicious cycle. Breaking this cycle is one of the most difficult challenges that confronts anyone concerned with planning environmentally sound development projects or the conservation of natural resources in developing countries. The situation may vary among countries, but the pattern remains essentially the same in most developing nations of Asia, Africa, and Latin America.

In contrast to the situation in biological sciences, the systems of education and government in developing countries provide several teaching curricula and lucrative jobs in disciplines traditionally considered important for economic development. Consequently, young people are attracted to the more renumerative disciplines, and engineers, economists, and such are produced in large numbers each year. These people predominate the field of economic development on numbers alone, and, moreover, become the decision-makers of developing countries. Unfortunately, their decisions are based on an educational background that lacks any exposure to the principles of ecology. Hence their plans for development rarely reflect an appreciation of the importance of wildlife habitats to social and economic welfare. Those who are officially empowered to question the ill-conceived plans are the unsophisticated and poorly paid wildlife managers, but socially and educationally these people are no match for the decision-makers. This in reality is the primary reason why economic development in developing countries frequently takes place with the indiscriminate alteration of wildlife habitats. In order to rectify this situation, it is important to expose decision-makers of developing countries to the principles of ecology and make them advocates of ecologically sound development planning. This exposure should also extend to politicians, industrialists, and lay people, so that the science of applied ecology and economic development could be integrated without delay. The long overdue need for a large-scale scientific mobilization cannot be overemphasized.

PROPOSED PLAN OF ACTION

Given the inadequate teaching staff in developing countries, the urgent need for training should at first be largely satisfied by ecologists of developed countries. Also, such training should be a long-term effort and should continue until a relatively large number of developing-country nationals are themselves capable of training their countrymen. Only then can the cycle created by the current systems of education and government be effectively broken. The long-term training of a large number of people is most feasible through programs conducted within developing countries. Moreover, due to the widespread need for such training, in-country programs should be simultaneously undertaken in many developing countries. The task of teaching applied ecology over long periods and in several developing countries is well beyond the capability of any one person or institution. Therefore it is important that several developed-country institutions make a concerted effort to transfer their knowledge of applied ecology to developing countries. Without the participation of several institutions it would be almost impossible to adequately address the current training needs of developing countries.

In-country Training Programs

In-country programs for teaching applied ecology should be established at wildlife management agencies, so that in-service personnel could be trained immediately. Also, it is necessary to institutionalize the teaching of applied ecology at local universities for at least three important reasons. First, unlike wildlife management agencies, universities cater to numerous disciplines including those that are normally considered crucial for economic development. Thus, institutionalizing applied ecology at universities creates opportunities for establishing interdisciplinary programs and promoting environmental awareness among future engineers, economists, and decision-makers. It also ensures that future biology teachers will have the knowledge and confidence to teach applied ecology at secondary and lower levels of education. Such teachers have enormous potential to promote environmental awareness in a developing country because they deal with the education of very large numbers of young people. The second reason for institutionalizing applied ecology curricula at universities is that university education has very high social value in developing countries. Hence a graduate with a knowledge of applied ecology can be effective in bringing about social change within communities. Third, universities are the only institutions that can confer postgraduate degrees, which are prerequisites for teachers of applied ecology at a university or a wildlife management agency. In most developing countries it is essential for such teachers to possess a doctorate degree, because a person with a doctorate not only brings respectability to the science of applied ecology but also moves into a special social class that can command relatively high wages and can effectively communicate with decision-makers. Thus, from both a social and educational perspective, it is important to institutionalize the teaching of applied ecology at universities. However, it may take at least a decade to institutionalize graduate and postgraduate curricula and correct the imbalances created by the current systems of education and government. Therefore, steps to institutionalize applied ecology both at universities and wildlife management agencies should be taken immediately.

The institutionalization of applied ecology in developing countries clearly requires a long-term and immediate commitment from individuals and institutions in the developed world. Janzen (1986), assessing the future of tropical ecology, points out the priority in this area: "Set aside your random research and devote your life to activities that will bring the world to understand that tropical nature is an integral part of human life." One of these activities is surely the teaching of applied ecology to nationals of developing countries. In fact, at present, a large number of well-qualified ecologists have field experience in developing countries, but do not have jobs in the developed world. These people would make ideal candidates for an Environmental Corps that could be developed along lines similar to the well-known Peace Corps. Developing countries are familiar with Peace Corps personnel and numerous types of consultants from the developed world. Therefore, accepting the ecologists of the Environmental Corps should pose no problems to them. The establishment of an Environmental Corps would be an important step in the transfer of ecological expertise from developed countries to developing countries.

Problems in Transferring Knowledge

The cross-cultural transfer of knowledge is not without problems, particularly with respect to the interpretation of concepts in applied ecology and wildlife management. For instance, in developed countries a national park denotes a relatively large area, free of any human exploitation of natural resources. This concept may not always be useful in developing countries because human survival in these countries has long depended on the exploitation of local natural resources. In fact, national parks, which were established using the developed world's concept, have frequently caused social problems in developing countries. In order to avoid such problems the national park concept must be taught in the context of local patterns of human exploitation of natural resources. Numerous other developed-world concepts and wildlife management practices—such as buffer zones, sustained yield management, and habitat enrichment—pose problems in interpretation and should be taught in the context of the local conditions in developing countries. Thus a teacher from the developed world should be knowledgeable in both the science and the social situation in the developing country in which he or she teaches.

Another problem in the cross-cultural transfer of knowledge is the language barrier. Only rarely can ecologists of developed countries be expected to use the language of a developing country as the medium of instruction. One way to partly circumvent the language problem is to include a large measure of practical work in the curriculum for teaching applied ecology. What is unclear during lectures becomes more easily understood when practiced in the field. Another way to partially overcome the language barrier is for ecologists of the developed world to gain fluency in more than one Western language such as English, French, Spanish, or Portuguese, which are still fairly widely spoken in developing countries. It is doubtful that the language problem can be entirely overcome unless and until applied ecologists of the developed world help to translate the scientific literature into the languages of developing countries. Therefore anyone who is committed to long-term teaching of applied ecology in developing countries should also be encouraged to learn at least some of the languages understood in these countries.

Public Education

In addition to professional training, in-country programs should be established for the education of nonprofessionals and the lay public to foster widespread appreciation of environmental conservation. At present, such programs in developing countries are mainly conducted by private organizations. Nevertheless applied ecologists of the developed world can improve these programs considerably by putting them on a more scientific footing. For instance, developed-country ecologists could help to explain the decline of a species on the basis of its ecology or the local environmental situation. Or they could explain an instance of human-wildlife conflict on the basis of wildlife ecology and human activities. Understanding the problem—rather than stating it, which is what most private environmental organizations in developing countries do—goes a long way in resolving the problem. Therefore collaboration between private environmental organizations and developed-country ecologists is important. Together they should organize a variety of activities such as seminars, exhibitions, and workshops to promote environmental awareness among the general public and the rural people. Such activities should also be organized for the education of decision-makers, particularly to apprise them of the negative impacts of specific development projects. A sense of professionalism could also be instilled into these programs through the participation of environmental education specialists from developed countries. The educational methods used by these specialists are somewhat different from those of scientists, and such specialists have been known to develop innovative environmental conservation programs even on the basis of culture and religion in developing countries. Therefore, it would be useful to include such people in the proposed Environmental Corps.

Overseas Training

In-country programs alone may not be sufficient to address the urgent need for training developing-country nationals in applied ecology. At present, a few developing-country university graduates have stayed in the biological sciences despite a bleak future and adverse financial consequences. Encouraging their manifest interest cannot be delayed until in-country curricula are established for the teaching of applied ecology. These people should be given the opportunity to enroll for a higher education at universities in developed countries. In developing countries, an overseas education has even more prestige value than education at a local university. Therefore, those nationals with overseas qualifications will enhance the process of social acceptance of applied ecology in their home countries and also accelerate the institutionalization of applied ecology at local universities and wildlife management agencies.

In the United States, nearly 150 universities offer undergraduate and graduate courses in wildlife management. Of these, about 90 universities have higher degree programs oriented toward natural resources management issues that foreign students may face in their own countries. While a large number of U.S. universities provide training in applied ecology, relatively few developing-country nationals have been able to avail themselves of this opportunity. The major constraint has been the lack of funds for fellowships, scholarships, and assistantships.

Governments of developing countries can ill afford the funds to educate their nationals in developed countries, since they have other, higher-priority needs to satisfy. Consequently, they often look to developed countries for financial assistance. Universities in developed countries have assisted with funds because their faculty members realize the importance of having foreign colleagues with similar interests. Private organizations and governments of developed countries have also provided funds, but the total assistance given by all of these institutions for teaching applied ecology to developing-country nationals still remains relatively small. It is time that international economic-development agencies also assist with substantial funding for this effort, especially because applied ecology is directly relevant and important for their activities in developing countries. These agencies should fund both in-country and overseas programs, and continue this assistance for as long as necessary.

The assistance of international economic-development agencies is also crucial for reasons other than financial support. These agencies have the capacity and leverage to persuade developing-country governments to pay attention to environmental conservation and education during the process of economic development. Consequently, they can and do bring about gradual social change in developing countries. Without initiating this change at the highest political levels, scientists will be hard put to teach or institutionalize the science of applied ecology in developing countries. Thus international economic-development agencies should initiate the proposed educational reforms and also provide the necessary funds to implement these reforms. Scientists will be more than happy to help in this venture.

Employment Opportunities

An important additional ingredient is needed to make the above-mentioned changes work. The newly trained cadre of developing-country ecologists will need employment. As indicated earlier, there are at present relatively few jobs for qualified ecologists in developing countries; this must be remedied by the creation of employment opportunities relevant to wildlife conservation and economic development. These opportunities will vary among countries because they depend largely on the specific conservation and development needs of each country. Hence a detailed plan to create these employment opportunities should be developed for each country. It would be unrealistic to propose a single strategy to resolve this difficult issue for all developing countries. However, we believe that such a strategy for any country should not gratuitously create a number of bureaus with weak connections to the mainstream of decision-making. It would be more meaningful to create jobs for ecologists within major decision-making agencies and within new and existing wildlife agencies that are included in the process of economic development. In creating these job opportunities we are convinced that the leverage of international assistance agencies and their collaboration with developing-country governments would again be crucial.

ONGOING TRAINING PROGRAMS

At present there are six well-known regional programs for training nationals of developing countries in applied ecology and natural resources management (Table 1). The oldest of these programs was established in Costa Rica in 1952, and the most recent one was inaugurated in India in 1982. All six regional programs emphasize fieldwork and focus on the training of middle-range wildlife managers. Consequently, all programs, except the one in Costa Rica, provide a training curriculum that leads to a wildlife-management certificate or a diploma. On the other hand, the Costa Rican program, which is affiliated with the country's National University, has a curriculum leading to a master's degree. It would be useful for the other five programs, as well, to be affiliated with local universities and to provide opportunities for appropriately qualified wildlife managers to obtain postgraduate degrees. The university affiliation is particularly important, at least until local universities have established their own curricula for training in applied ecology. Without this affiliation present-day wildlife managers will lose the opportunity to be assimilated into the university educational system, which has high social value.

The duration of the curricula for wildlife-management certificates and diplomas differs considerably among regional programs. In the Indonesian training program, the diploma takes nine months. In the Indian and the two African programs, diploma courses last two years. The Indian and African programs also provide courses for wildlife management certificates that take anywhere from four and a half months to one year. The master's degree in the Costa Rican program takes two to three years, including one year for the completion of

TABLE 1

REGIONAL SCHOOLS FOR TEACHING APPLIED ECOLOGY TO
DEVELOPING-COUNTRY NATIONALS

Location	Year Established	No. of Participants (Countries/Individuals)	Type of Curriculum	Duration
Mweka, Tanzania	1963	16/1000	Wildlife Mgt. Certificate Postgrad. Diploma	1 year 2 years
Garcua, Cameroon	1970	20/400	Wildlife Mgt. Certificate Postgrad. Diploma	1 year 2 years
Bogor, Indonesia	1979	9/200	Wildlife Mgt. Diploma	9 months
Dehra Dun, India	1979	«/300	Short Courses Wildlife Mgt. Certificate	1–2 weeks $4^{1}/_{2}$–10 months
Bariloche, Argentina	1968	3/– 5/–	Postgrad. Diploma Wildlife Mgt.	2 years 1 year Diploma
Turrialba, Costa Rica	1952	10/200	Short Courses M.S. Degree	1–3 weeks 2–3 years

SOURCES: Budowski, 1986, and Thorsell, 1986.

prerequisites. In addition to the above-mentioned curricula the Indian and the Costa Rican programs offer short-term courses focusing on specialized wildlife-management techniques or on specific target groups such as planners and decision-makers. In the future, these two programs are also expected to develop management-oriented research projects, disseminate scientific information, and provide wildlife-management consultancies to the countries of the region.

In all, these six programs cater to the training needs of as many as 60 tropical countries. This situation has led to some concern and discontent that at least a few of these programs emphasize local training needs rather than regional requirements. Instead of the current average of one program for 10 countries, the average should be more like 10 well-funded programs in each developing country, considering the urgency for wildlife training and the importance of promoting the social acceptance of environmental conservation. Of these programs, a few could cater to the professional training of wildlife managers. The others must address the need for institutionalizing applied ecology curricula at universities, the education of nonprofessionals, including lay people, and the overseas training of wildlife managers and biology graduates. By implementing these programs, the existing situation in wildlife-management training and environmental conservation can be greatly improved in developing countries.

Three other programs should be mentioned with regard to training in applied ecology for developing-country nationals. One is the interdisciplinary program in ecology established by the Indian Institute of Science in Bangalore. This program engages in staff research projects, organization of seminars and symposia, and the training of doctoral candidates in ecology. In addition, this program has conducted a six-month training course in wildlife biology for participants with advanced qualifications. The other two programs are based in Great Britain and the United States, respectively. The program in Great Britain is affiliated with the University of London and offers a one-year postgraduate course in conservation. The U.S. program is sponsored by the Smithsonian Institution and offers short-term courses as well as long-term support for the training of developing-country wildlife managers.

The Smithsonian program, based at the National Zoological Park in Washington, D.C., was inaugurated in 1981. This program provides short-term training courses of five to ten weeks that are conducted overseas as well as in the United States, at the National Zoo's Conservation and Research Center, Front Royal, Virginia. The U.S. courses, which are held during the summer, cater to foreign students based at U.S. universities, and to individuals residing in tropical countries who wish to follow advanced training in wildlife management. In addition to wildlife-management training, the U.S. courses include curricula for captive-animal management and conservation education, which will be expanded in the near future. The training is provided through lectures, seminars, workshops, and substantial amounts of fieldwork leading to the completion of short-term research projects. The curriculum for the overseas courses is similar, but when necessary, these courses focus on and help resolve local wildlife problems through the development of field projects. Overseas courses have been conducted in seven countries (Argentina, Brazil, Peru, Venezuela, Malaysia, Indonesia, and Sri Lanka) so far. These courses have also been repeated in Venezuela and Sri Lanka in order to monitor previously established field projects, transfer new technology, or train more people. Thus the short-term overseas courses have gradually transformed into long-term support for wildlife management and training in at least two countries. Long-term support of this nature will be also given to Malaysia and Peru in 1987 and 1988, respectively.

Other aspects of the long-term support given by the Smithsonian program include assistance for the institutionalization of training courses at overseas agencies and for the postgraduate education of former course participants. The first-mentioned aspect was initiated only in 1986 and therefore is still in the developmental phase. Nevertheless steps have been taken to institutionalize wildlife-management training curricula in Malaysia and India. These steps include the training of former course participants as instructors and the development of curricula, specially designed for the training needs of each country. In Malaysia the curriculum will be established at the West Malaysian Department of Wildlife, while in India it will be institutionalized through the University of Mysore. The Malaysian agency has indicated its desire to institutionalize the training as a regional program for Association of South East Asian Nations (ASEAN) countries. Assistance for the higher education of former course participants was made possible through collaboration with some U.S. universities, especially through a joint venture with the University of Florida, Gainesville. After the short-term training courses nearly 15 participants have been assisted with placements or financial support for master's and doctoral programs at five U.S. universities. Also, the Smithsonian program has financially supported two Ph.D. candidates of Indian universities to complete their dissertation research in the United States. Some recipients of assistance for postgraduate education are currently active in wildlife projects in their home countries, while others are still continuing their higher-degree programs in the United States.

Between 1981 and 1986, 199 wildlife biologists and conservationists from 21 countries were trained in the 16 field courses conducted by the Smithsonian program. This includes 102 people from 10 Asian countries, 86 from 5 South American countries, and only 11 from 5 African countries. Clearly, not much headway has been made in Africa, but this is mainly due to the lack of funds to hire additional instructors for the Smithsonian program. Until the beginning of 1987 the program included only one full-time instructor. Now, there are two instructors and plans are under way to conduct courses next year in at least one of five African countries that has requested in-country training. Nearly two-thirds of the trainees have been employees of overseas wildlife-management agencies while 21 percent were affiliated with universities and 11 percent belonged to nongovernmental organizations. Through this broad-based institutional participation the Smithsonian program has promoted the general principle that, while professional training is important, everyone who can support wildlife conservation and management in developing countries should have a knowledge of applied ecology. About three-fourths of the trainees had at least basic degrees, but the program has also trained wildlife managers with considerable field experience.

The Smithsonian program is now in the process of developing a new curriculum for a potentially significant but neglected educational institution in the developing world—the zoological garden. As in Western nations, the Third World zoo is an extremely popular setting for family outings. Consequently it is a place where adults and children can learn to appreciate nature and wildlife. This potential, however, is mitigated by the poor conditions of most zoos. The new training curriculum is targeted for mid-level zoo managers and veterinarians, with the object of enhancing the care and management of wild animals in captivity. Improving the management of captive animals can foster an appreciation of and respect for wildlife, public education concerning environmental and natural resources issues affecting wildlife, and viable captive breeding programs for endangered species. The first two courses began in Thailand and Malaysia in late 1987 and early 1988, respectively.

In conclusion, it should be stated that the Smithsonian program has initiated many aspects of the plan of action proposed earlier. Nevertheless, almost all aspects of the program must be greatly strengthened and expanded in the future. This could be achieved through recruitment of more teaching staff and with financial support for the activities undertaken by the Smithsonian program. In other respects the Smithsonian Institution is well equipped to handle a comprehensive program for the training of developing-country nationals in applied ecology. In 1988, a large building complex at the 3,000-acre facility at Front Royal is slated for renovation for the establishment of an international wildlife training center.

References Cited and Sources of Further Information

Abrahamson, D., 1983. What Africans think about African wildlife. Internat. Wldlf. 13:38–41.

Allen, R., 1980. The world conservation strategy: what it is and what it means for parks. Parks 5(2):1–5.

Blower, J., 1984. Terrestrial parks for developing countries. *In:* J. A. McNeely and K.R. Miller (eds.), National parks, conservation and development: the role of protected areas in sustaining society. Washington, DC: Smithsonian Institution Press, pp. 722–727

Bryant, J., 1983. Conservation directory—a list of organizations, agencies and officials concerned with natural resource use and management. Washington, DC: National Wildlife Federation, 301 pp.

Budowski, G., 1986. Training: CATIE shows how in Central America. IUCN Bull. 17:35–36.

Council on Environmental Quality (CEQ) and the U.S. Department of State (USDOS), 1980. Global 2000 report to the President. Washington, DC: Government Printing Office, 3 vols.

Commoner, B., 1966. Science and survival. New York: Viking Press, 150 pp.

Dasmann, R.F., 1975. National parks, nature conservation and future primitive. *In:* Proceedings of the South Pacific conference on national parks and reserves. February 24–27, Wellington, New Zealand, pp. 89–97.

Dietz, L.A., and E. Nagagata, 1986. Community conservation education program for the golden lion tamarin. *In:* Case study summaries of the workshop on building support for conservation in rural areas, May 27–31, Highgate Springs, VT, pp. 5–13.

Ehrlich, P.R., 1968. The population bomb. New York: Ballantine Books, 223 pp.

International Union for Conservation of Nature and Natural Resources

(IUCN), 1975. The world directory of national parks and other protected areas. Gland, Switzerland: IUCN, vol. 1.

IUCN, 1986. IUCN and education: laying the foundation for sustainable development. IUCN Bull. 17:112–125.

IUCN, U.N. Environment Programme (UNEP) and World Wildlife Federation (WWF), 1980. World conservation strategy: living resource conservation for sustainable development. Morges, Switzerland: IUCN, 54 pp.

Janzen, D.H., 1986. The future of tropical ecology. Ann. Rev. Ecol. Syst. 17:305–324.

Kelley, R., 1985. Natural resources and environmental management at North American universities—a guide to training opportunities. Washington, DC: World Wildlife Fund—U.S., Rare, International Institute for Environment and Development and U.S. Agency for International Development, 306 pp.

Lanly, J.P., 1982. Tropical forest resources, Forestry Paper No. 30. Rome: Food and Agriculture Organisation, 106 pp.

Machlis, G.E., and D.L. Tichnell, 1985. The state of the world's parks—an international assessment for resource management, policy and research. Boulder, CO: Westview Press, 131 pp.

Marks, S.A., 1984. The imperial lion: human dimensions of wildlife management in Central Africa. Boulder, CO: Westview Press, 196 pp.

Medawar, P.B., 1959. The future of man. New York: Basic Books, 128 pp.

Mishra, H.R., 1982. Balancing human needs and conservation in Nepal's Royal Chitwan Park. Ambio 11:246–251.

Myers, N., 1979. The sinking ark. New York: Pergamon Press, 307 pp.

Myers, N., 1980. Conversion of tropical moist forests. Washington, DC: National Academy of Sciences, 205 pp.

Myers, N., 1984. Too big for their own good? Internat. Wldlf. 14:26–32.

Ramade, F., 1981. Ecology of natural resources. New York: Wiley and Sons, 231 pp.

Saenger, P., E.J. Hegerl, and J.D.S. Davie, 1983. Global status of mangrove ecosystems. Environmentalist 3(3):1–88.

Swaminathan, M.S., 1986. Education for a viable future. IUCN Bull. 17:111.

Thorsell, J., 1986. We need the managers. IUCN Bull. 17:126–127.

Train, R.E., 1984. Biological diversity: ecological basis for sustainable development. J. World Resources Institute, pp. 27–33.

U.S. Interagency Task Force, 1985. U.S. strategy on the conservation of biological diversity: report to congress. Washington, DC: Government Printing Office.

Wemmer, C.M., J.L.D. Smith, and H.R. Mishra, in press. Tigers in the wild: the biopolitical challenges. *In:* R. Tilson and U.S. Seal (eds.), Tigers of the world. Parkridge, NJ: Noyes Publications.

Chapter 7

INTEGRATION OF BIOLOGICAL CONSERVATION WITH DEVELOPMENT POLICY: THE ROLE OF ECOLOGICAL ANALYSIS

KATHRYN A. SATERSON

Although the term "biological diversity" is quickly becoming familiar to scientists, policy-makers, and even the general public, it does not yet have a generally accepted definition and must be defined by each user. For ecologists and development agencies, biological diversity is best defined as including the genetic diversity within species, the number of different species in an area, and the variety of ecosystems. Another definition is the variety and variability among living organisms and the ecological complexes in which they occur. Perhaps the *World Conservation Strategy* (IUCN, 1980) provides the simplest definition of biological diversity by referring to living resources. Living resource conservation includes preserving genetic diversity and ensuring the sustainable use of species and ecosystems.

A conservative estimate of the current rate of destruction of tropical

141

evergreen forests due to clearcutting, shifting agriculture, and fuelwood gathering is about 80,000 square miles per year. Even if we use the most conservative estimates of current diversity and rates of habitat alteration and extinction, conservation of biological diversity is a global problem that affects all species, including humans. While ecologists and conservationists have become increasingly aware over the last 15 years of the ecological, economic, and aesthetic importance of conserving biological diversity, this awareness has only recently begun to be acknowledged in the programs of the international development community.

During the 1980s, multilateral and bilateral donor institutions and the governments of developing nations have increasingly recognized that sustainable economic development depends directly on the conservation and sustainable use of natural resources. Slower in emerging has been the recognition that the biological diversity contained in both domestic and wild plant and animal resources is also important for sustainable development and must be maintained not only in protected areas, but also within agricultural, range, marine, and forest systems being managed for their biological productivity.

Two events in development institutions reflect the emerging recognition that conservation of biological diversity is related to economic development and human welfare. Governments of other nations are also recognizing this relationship. In his address to the National Forum on Biodiversity in September 1986, Dr. Nyle Brady, senior assistant administrator for science and technology with the Agency for International Development (AID), spoke on "International Development and the Protection of Biological Diversity." Dr. Brady noted that "sustained economic development requires conservation of biological resources." and that to meet this challenge we must find ways to "conserve more natural habitats, better manage those that already exist . . . and increase food and fuelwood production on land already cleared in order to reduce pressure on remaining wild areas" (Brady, 1986).

Dr. Brady also acknowledged that conservation of biological diversity is essential to maintaining ecological services such as protection and regulation of soil and water resources, cycling of nutrients, and buffering of climate extremes. In 1987, AID responded to new congressional legislation by obligating 2.38 million dollars for more than 30 projects directly related to conservation of biological diversity.

Another significant event was the July 1986 World Bank adoption of an operational policy on wildlands. This policy directs the World Bank to finance projects on lands that are already disturbed and to mitigate use of undisturbed lands by financing preservation of ecologically similar areas.

Despite advances such as these, scientists still must investigate the ecological principles that control the maintenance of biological diversity, for we cannot yet fully predict the impact of human activity on biological diversity. In addition, ecologists must continue to apply ecological data, theories, and perspectives to development projects and to demonstrate the ways that maintenance of biological diversity contributes to sustainable economic development.

BASIC ECOLOGICAL THEORIES AND ANALYSES RELATED TO BIOLOGICAL DIVERSITY

While systematists are focusing on species counts for a number of habitats around the world, ecologists are attempting to describe some of the factors and mechanisms that control species, habitat, and ecosystem diversity. However, the controlling factors vary greatly among communities and ecosystems. Correlations between terrestrial diversity and maritime influence, precipitation, grazing intensity, and soil fertility are not consistent. Even when correlations such as a latitudinal diversity gradient are observed (biotic diversity decreases from the Equator toward both poles), the causes for the correlation are still debated. A number of biological mechanisms (niche relations, habitat diversity, mass effects, and ecological equivalency) can contribute to the maintenance of diversity at the species and ecosystem level.

In addition to addressing the mechanisms that govern the creation and maintenance of biological diversity, ecologists have also examined the contribution of biological diversity to the functioning of ecosystems. Kimmerer (1984) suggests "dropping the diversity-stability relationship from the ecologists' repertoire" because some ecosystems show an increase in stability and resilience with an increase in biological diversity, and others do not (Johnson, 1986). However, other scientists still hold that additional research is needed on the relationship between diversity and ecosystem stability and resilience.

Research on the effects of habitat disturbance on diversity does not reveal a simple positive relationship between increased disturbance and decreased diversity. In many systems, low to moderate levels of disturbance may result in increased diversity, although this may reflect an increase in the diversity of exotic species at the expense of a decrease in native species. Severe disturbance may lead to less diverse communities where native species are replaced by exotic. In a forest in Uganda light to moderate disturbance (25 percent harvesting destruction) caused only temporary disruption of natural tree-fall rates whereas heavy disturbance (50 percent destruction) seriously increased natural tree-fall rates. Although 35 percent destruction is the maximum disturbance that allows maintenance of a healthy forest mosaic, this conclusion could be questioned on the basis that there could be significant changes in the forest's composition or functions before a change in natural tree-fall rates is observed.

Theories of island biogeography have been called upon, with varying degrees of success, to explain the distribution and maintenance of biological diversity. The theory of island biogeography is based on the observation of a general logarithmic relationship between the number of species and the size of an island (both true islands and islands of isolated habitat), as well as the distance of the island from the mainland or source of gone flow. The theory predicts that an area reduced to 10 percent of its original size would support only half the original species present. While this conceptual framework is useful, we must—in addition to determining the minimum areas needed to preserve existing species—determine the minimum areas needed to preserve existing ecological and evolutionary processes. Theories of island biogeography can be tested in areas where natural or human activity has reduced habitat.

The World Wildlife Fund and Brazil's National Institute for Amazon Research are conducting an exciting, long-term study of the effects of habitat isolation on diversity and other processes as rain forests of the Brazilian Amazon are cleared to create pasture. In order to determine the minimum critical size required to maintain the original forest diversity, the project involves study of the changes in diversity in forest fragments of 1, 10, 100, 1,000 and 10,000 hectares. Initial results indicate that there was a loss of bird species and an increase in tree mortality rates in isolated 1- and 10-hectare reserves. The results and insights emerging from this project will be useful in testing island biogeographic theory, providing design and management recommendations for protected areas in the Amazon, and providing basic information on the degree of species composition structure in natural communities.

Although the Minimum Critical Size Project is providing much needed empirical data, the fragment sizes are relatively small and it could take decades before conclusions can be drawn about changes in the largest forest fragments. Theoretical minimum-viable-population analyses are currently used to test theories and generalizations on a much larger scale. They also can be used to identify and estimate the influence of natural processes on species losses.

The concept of a minimum viable population of a species is associated with some difficulties but it responds to our pressing need for conservationists to develop a predictive understanding of the relationship between a population's size and its chances of extinction. Opportunities for reserving significant quantities of habitat are disappearing so rapidly that we cannot rely solely on long-term studies of biogeographic equilibrium theory to provide insights into the controls of species occurrence and persistence.

Need for Additional Research

Although ecologists must begin to apply present knowledge to the conservation of biological diversity, the lack of scientific consensus on many theories indicates the need for continued research in a number of critical areas. Research needs include:

- the effects of habitat fragmentation on diversity, in order to predict species losses and ecosystem degradation as undisturbed habitats become increasingly smaller and isolated;
- mechanisms controlling species, habitat, and ecosystem survival at both the local and regional scale;
- methods to estimate the value of conserving species and ecosystems in order to develop conservation priorities;
- what is the most appropriate size of a protected area in different ecosystems, and how essential are dispersal pathways or corridors linking areas;
- the importance of biological diversity to systems with intensive human use, such as agricultural and marine systems;
- the potential for sustained yields of useful animal and plant species from natural systems—natural forest management is an area worthy of much more research;

- rehabilitation of lands to restore biological diversity and normative ecological processes.

APPLICATION OF ECOLOGICAL ANALYSES TO SUCCESSFUL DEVELOPMENT AND CONSERVATION PROJECTS AND POLICIES

Ecologists can contribute their expertise at technical and political levels, both in terms of designing actual field projects and developing institutional policies. There are some exciting examples of the role ecological analyses have played in development projects. Unfortunately, sound scientific analyses of conservation priorities that can be applied to development projects are lacking, and such priorities considered in light of nonscientific (political, economic, and social) constraints are almost completely absent.

Successful conservation of biological diversity in the development context depends not only on a sound scientific basis, but also on an acknowledgment of the economic, political, and social context. Because economic development is the primary goal of less-developed countries and development agencies, the conservation of many biological resources will depend on the demonstration of their economic value.

The development of appropriate and successful policies for sustainable use and conservation of natural resources depends on the contributions of ecologists who can effectively transfer their knowledge of basic science to nonscientists, understand the nonscientific constraints to managing resources in developing nations, and work collaboratively with social and agricultural scientists.

Local, national, and international policies that affect the conservation of biological diversity can benefit from the contributions of ecologists and biologists. Ecologists who transferred their knowledge of biological resources conservation to the political arena were influential in bringing biological diversity and tropical forest conservation to the attention of the U.S. Congress in 1986. Letters from members of the Ecological Society of America and the American Institute of Biological Sciences to their congressional representatives encouraged passage of the 1986 amendments to the Foreign Assistance Act. Those amendments required that AID spend 2.5 million dollars on biological diversity projects. Ecologists can be influential in making Congress and development agencies aware of the need to integrate biological conservation with broader natural-resources management projects.

Because it is difficult to generalize about what controls diversity, it is important, while encouraging policy-makers to address ecological issues, to also make them aware of the limitations of current information. It is often inappropriate to extrapolate data from temperate regions to the tropics and it is even difficult to be certain that data from one tropical area can be applied to another. Scientists must meet the challenge of making complex theories and data understandable to policy-makers without creating inaccurate oversimplifications.

Protected Areas: Identification, Design, and Management

IDENTIFICATION. Ideally, we would like samples of all species, communities, and ecosystems to be maintained in protected areas. Unfortunately, the realities of our current limited knowledge of species and habitat requirements, combined with existing development pressures, indicate that we must identify habitats most in need of immediate protection. Acknowledging the limitations of ecological knowledge, it is nevertheless necessary for ecologists to help identify areas worth protecting (due to high concentrations of rare or endemic species, or provision of essential environmental services).

A number of criteria must be considered in evaluating the need to protect biological diversity in a given geographic area. Included in ecological, administrative, and economic criteria are:

- level of species endemism and habitat richness;
- degree of human threat to that diversity and the intrinsic vulnerability of the species or ecosystem;
- importance (both economic and ecological) of the species or ecosystem to the human needs of the area;
- importance of the habitat to maintaining diversity in other regions;
- amount of similar habitat protected elsewhere; and
- political and economic conditions that could encourage or hinder successful protection.

The 15 criteria that should be considered when judging the conservation value of particular areas include: size, richness and diversity, fragility, and opportunities for conservation. Currently, major conservation areas worldwide number 3,500 (from nature reserves to multiple-use natural resources lands), totaling about 4.25 million square kilometers. Although these areas represent about 182 of the 193 biogeographical provinces recognized by the International Union for Conservation of Nature and Natural Resources (IUCN), many habitat types are missing, and many are represented by areas that may be too small for adequate protection.

In the mid-1970s, the Brazilian Amazon Conservation Plan gave first priority to areas that two or more scientists identified as possible Pleistocene refugia (areas likely to support many endemic plants and animals). Scientists with a number of nongovernmental organizations in the United States are attempting to describe the countries with the greatest need for conservation of biological diversity. In those cases, ecologists are contributing to the definition of technical criteria for conservation of biodiversity. They must then interact with policymakers to determine the weight that technical criteria should receive, in relation to political, economic and administrative criteria.

DESIGN. Once areas in need of protection are identified, ecologists can then assist in determining the minimum area needed to sustain the floral and faunal populations in light of the specific anthropogenic pressures they face. Theories of island biogeography and the results of projects like the minimum critical area project in the Amazon will be useful in predicting what size of fragmented habitat is necessary to ensure the survival of certain species. MacKinnon et al. (1986) outlines guidelines based on island biogeography. The guidelines include making areas as large as possible, including year-round habitats of native species, encompassing a wide continuous range of ecological communities, and making sure areas are not completely isolated from other protected areas without corridors for dispersal. The choice between one large reserve or many smaller ones depends on the particular conservation objective and the actual biology of the situation. Edge effects due to forest isolation in Brazil imply that reserves must include large buffer zones ". . . where edge-related changes can take place without affecting the large stable core areas." (Lovejoy et al. 1986).

Other useful approaches to determining the appropriate size of protected areas give less emphasis to empirical observations. (Soulé, 1987). Analyses based on minimum viable populations of species (Belovsky, 1987; Shaffer, 1985, 1987) indicate that the minimum area required to adequately conserve some species can be substantial. The rarest species or those most subject to catastrophes or environmental variability are the species that would require the largest population size (or largest reserve) to ensure a good chance of long-term persistence; multiple populations of adequate size are recommended for virtually all species.

MANAGEMENT. Ecologists can also give insights into how to integrate the maintenance of floral and faunal populations with human land-uses. Many protected areas lack adequate management and in some cases intensive human management may be necessary to reintroduce species to restore diversity. Minimum-viable-population analyses can demonstrate how available information can be used to assess the potential long-term consequences of short-term management decisions. MacKinnon et al. (1986) summarize many of the important management considerations in *Managing Protected Areas in the Tropics*.

Conservation Outside Protected Areas

Maintenance of biological diversity depends on understanding the ecological and socioeconomic effects of major forest, hydropower, and agricultural development projects; of human activities as diverse as clearcutting forests, introducing foreign species, eliminating predators, dynamite-fishing, and using pesticides; and of global phenomena resulting from industrialization, such as acid rain and climate change. All these activities must be analyzed ecologically if we are to mitigate their local and global effects on biological resources.

It is important that ecologists be included on the design teams for development projects to assess likely impacts on biological resources as projects are being designed, rather than doing environmental assessments after substantial commitments of time and energy have gone into project planning. Environmental assessments are increasingly being conducted simultaneously with project designs, at development agencies like AID. AID recently included an ecologist on the design team for an agricultural development project in Bolivia. This resulted in the successful incorporation of ecological analyses into basic project design; irrigation plans were changed, a road plan was altered, and recommendations for protected areas were made.

The lack of appropriate monitoring and evaluation of natural resources impacts in most development projects means that we are ignoring valuable information that could be useful in improving future development projects. An environmental impact statement should be viewed as a scientific prediction and monitoring should be designed to test specific theories about environmental processes. Important ecological lessons can be learned, for example, from observation of the effects of psyllid aphids on *Leucaena* monocultures in Asia. Ecologists can inform policy-makers of the need to monitor environmental impacts of projects, and recommend approaches and appropriate time scales. The three- to six-year time-frames of many development projects dealing with natural resources may not be adequate to yield ecological information on project benefits and sustainability. Much-longer-term monitoring adds considerably to project expense but is necessary to test for sustainability.

Development agencies are looking to groups such as the Conservation Data Centers, established by the Nature Conservancy, to provide basic biological information to assist in designing environmentally sound projects. However, scientists with these groups must learn to provide information in ways that are useful to development planners, and to produce socially, economically, and politically realistic recommendations. Unfortunately, preservation of all threatened habitats and species is rarely an option, and difficult choices must be made.

Most ecological theories on the structure, function, and maintenance of ecosystems recognize the structuring role of cyclical natural disturbances such as fire, pests, and floods. In order to design development projects that maintain biological diversity in nations under intense human pressure, we must examine ecological theories for their applicability to human disturbances. One strategy for in situ preservation of crop and wild plant genetic resources is maintaining traditional farming systems (diverse polycultures) and the adjacent natural ecosystems. Ecological analysis of the ways that traditional farm systems maintain crop and wild plant resources can contribute to agricultural development policies. Forest reserves could be created to improve the quality of local agriculture by maintaining biological diversity.

In many humid tropical areas the principle threat to conservation of biological resources is habitat alteration caused by agricultural encroachment by landless people and by larger development projects. Agricultural practices alternative to slash-and-burn systems are urgently needed in areas where human population density is high or rapidly increasing. Ecologists could contribute to promising agricultural development work related to multipurpose tree species and use of alley cropping of trees with annual food crops. Systems analyses of agroeco-systems have been used to define research and development priorities for improving food production by small farmers on marginal land. Ecologists can also identify areas where sustainable agricultural production may be unrealistic, such as on some steep, nutrient-poor slopes.

Specific Fields of Ecology and Project Examples

Ecological data and theories from many fields must be applied to sustainable development projects that successfully incorporate the maintenance of biological resources. The following are a few examples of development projects that have benefited from ecological analyses in specific fields (although insights from more than one field of ecology have contributed to each project).

The work of population ecologists on particular species can be used to estimate the resistance and resilience of plant and animal populations to human exploitation or other stresses. The habitat requirements of species can assist in determining the type and amount of protected area that will be required to maintain a population. Nepal's Royal Chitwan National Park contains a significant population of more than 350 en-dangered, greater one-horned rhinoceros (*Rhinoceros unicornis*); nearby are a number of villages that contain more than 20,000 people who depend on the park resources for fuelwood, thatch for their homes, food, and other purposes. The Smithsonian Institution, AID, and the World Wildlife Fund are supporting basic research aimed at determin-ing the population dynamics of the rhinos and at estimating the effects of the villagers' use of resources on rhino habitat. A graduate student has studied the response of the park's grassland areas to harvesting and the amount of grass needed to sustain the local population that depends on the park. A new component of the project includes development of forest plantations outside the park in order to lessen the local villagers' need for resources from within the protected area.

Theories from the field of community ecology can provide insights into the consequences of actions that change diversity. Theories about the way plants respond to environmental changes in forest openings (gap-phase dynamics) due to windfall or fire must be applied to larger-scale gaps created by human activities such as harvesting or pollution. New knowledge about gap-phase dynamics and the regeneration requirements of tropical forest canopy-tree species formed the basis for a natural forest-management plan being tested as part of a major AID project in Peru. The Central Selva Resource Management Project, in the Palcazu Valley, is funded at 30 million dollars over six years by AID and the Peruvian government. The project aims to test and institutionalize a sustained production forestry method based on rotational, narrow clearcuts over a 30-year cycle, while protecting the traditional culture of the local Amuesha Indians.

The integrative approach of ecosystem ecology is useful in determining which representative systems need protection and in ensuring that development projects do not permanently change the regulation of major ecosystem processes. The Central Selva project also includes establishing the Yanachaga/Chemillen National Park, in part because of its value in protecting downstream production forestry and agriculture.

An awareness of nutrient cycling is essential for protected-area managers, agriculturalists, foresters, range managers, etc., to determine how to manage various natural communities (e.g., burning grasslands). The importance of wetlands and their influence on surrounding areas, in terms of nutrient cycling and flood control, must be considered in coastal resources management projects.

Insights from the relatively new field of restoration ecology can be used to restore the productivity of degraded lands in order to lessen the pressure to develop undisturbed habitats. This field of ecology might also be applied to re-creating communities, such as the tall-grass prairie in midwestern United States; to reforestation programs that encourage the use of a mix of native species, rather than monocultures of introduced species; and to the enhancement of natural regeneration. Daniel Janzen is attempting to develop Guanacaste National Park, a 700-square-kilometer area in northwestern Costa Rica, by rehabilitating many hundreds of hectares of degraded, dry tropical forest. This project will use fire control, natural seed dispersal, and reforestation to allow the plants and animals from the dry forest in nearby Santa Rosa National Park and from evergreen-forested volcano slopes to reoccupy adjacent, low-quality agricultural and pastoral land. Elsewhere, the work of physiological ecologists can contribute to increasing the productivity of degraded lands by identifying crops that are adapted to highly acidic, saline, infertile, or poorly drained marginal lands.

Tropical marine resources are not mentioned as frequently as rain forests in discussions of diversity, but they are extremely important biologically and economically. AID and the National Oceanic and Atmospheric Administration (NOAA) recently issued a report on opportunities for economic development and management of Caribbean nearshore marine-habitat resources. This report noted a need for coastal resources-management plans that integrate marine resource conservation (including the establishment of marine protected areas) with income generation. Marine ecologists can assist with such planning and required training of personnel. Marine park planning and management is also included in an AID coastal resources project in Phuket, Thailand.

INTEGRATION OF BIOLOGICAL DIVERSITY CONSERVATION INTO SUSTAINABLE DEVELOPMENT PROJECTS

The *World Conservation Strategy* (IUCN, 1980) discusses the relationship between natural resources management and sustainable development and outlines the many ways that biological diversity is important to human welfare (food, pharmaceuticals, etc.). Despite a number of noteworthy development projects that include biological resources conservation as a component, we still see the need to demonstrate more clearly that biological resources conservation (and not just broadly defined natural resources management) must be taken into account in all development projects that affect natural resources. In order for this to occur, we need more data and examples of how diversity affects the viability and functioning of different specific ecosystems. Science must provide more than intuitive answers to the question of whether conservation of biological diversity is a luxury or a necessity. What amount of biological diversity is necessary to conserve? The question of short-term economic gains versus long-term economic and biological losses must also be addressed.

Ecological analyses have the power to demonstrate to policy-makers and development institutions that conservation of biological diversity is a necessity, not just a luxury. The more often this is successfully demonstrated, the greater the likelihood that institutions such as the World Bank and AID will implement their innovative wildlands and resources-management policies. If the World Bank wildlands policy is fully implemented, it could go far toward integrating biotic resources conservation with economic development in a large number of humid tropical countries. However, the governments of less developed countries must be willing to borrow under the conditions of the wildlands policy, so the economists and financial planners must be convinced that such investments are worthwhile.

Integrating Biological Diversity with Other Resource Management Concerns

In many countries donors become involved with natural resources only when the issues are a part of location-specific development projects. Even when impacts on natural resources are considered, biological diversity may be ignored. Therefore, biological diversity is frequently considered only on an ad hoc basis, if at all. We need more region-wide proactive biological diversity policies and national conservation strategies. The AID overseas missions are now required to address biological diversity in their country-development strategies. Many missions solicit input from ecologists in the preparation of assessments of the actions necessary to conserve biological resources in the host country.

Half of the 219 AID activities related to conservation of biological diversity in 1986 were in the area of natural resources and environmental management, while the other half were in protection, enhanced production, and germplasm conservation. AID is supporting a ten-year natural resources project in Panama that includes watershed management and soil conservation, national parks and wildlands management, natural forest management, and development of private industrial plantations and woodlots. The project design acknowledges the need to protect the economic returns from commercial investments, such as the Panama Canal and existing agricultural projects. Protecting the watershed above the canal should lessen siltation into Lake Madden, and thus help protect the water and electricity supply for millions of people in Panama City. The natural diversity of the watershed offers a resistance to insect and disease outbreaks that a monoculture forestry plantation cannot.

Another AID project that integrates the conservation of a biological resource with broader natural resources-management concerns is the Mahaweli Environment Project in Sri Lanka. The government of Sri Lanka's Accelerated Mahaweli Development Program, begun in 1978, includes the construction of four major dams and the irrigation and agricultural development of 102,000 hectares of dry land. Although some initial assessments of environmental impacts were made by the government of Sri Lanka and AID as part of individual component feasibility studies, a comprehensive environmental assessment of the program was requested by the government of Sri Lanka, funded by AID and completed in 1980. The government of Sri Lanka incorporated the recommendations of the environmental assessment into an environmental plan of action for the Accelerated Mahaweli Development Program.

In 1982, AID helped the government of Sri Lanka to develop the Mahaweli Environment Project to ensure the stability of the irrigated agricultural development by protecting areas for displaced wildlife, primarily the endangered Asian elephant. The 6.9-million-dollar project budget included 5 million dollars as an AID grant. The project includes establishment of four national parks, development of buffer zones, strengthening of Department of Wildlife Conservation staff and management plans, and the development of Department of Wildlife Conservation training capability. The World Wildlife Fund, the International Union for Conservation of Nature and Natural Resources, and the Food and Agriculture Organisation have provided funds for wildlife surveys. AID and other donors have supported watershed management actions that include reforestation.

The conservation of biotic resources and the establishment of protected areas are directly useful for providing economic return and more indirectly (but no less importantly) for the ecological functions of erosion control, water recharge and purification, flood control, maintenance of genetic resources, and so on. Ecologists can assist with the education of decision-makers and local populations about the value of biological resources and the costs of not conserving them. They can also work with economists to demonstrate the economic cost/benefit of exploiting biological resources that are becoming irreplaceable. The economic contributions of wild plants and animals to developing countries have been studied as has the value of conserving genetic resources. These analyses discussed the economic benefits derived directly from resources such as fish, herbs, wood products, genes for pest resistance, mangrove ecosystems, and wildlife and, indirectly, from the environmental services that those resources provide.

Biological Diversity and Local Human Needs

Development agencies and all governments must develop policies that promote and ensure the conservation of natural resources while also maintaining the livelihood of local people. Many natural systems have been selectively harvested for plant and animal resources for hundreds of years at scales that are within those systems' abilities to regenerate naturally. However, increased population pressures and technical innovations often lead to more intensive use of lands and result in a decrease in a system's ability to meet the basic needs of local peoples.

In September of 1985, AID awarded a grant to the World Wildlife Fund to develop a field program in wildlands and human needs. This project seeks to integrate management of natural resources and preservation of biological diversity with the development needs of local human populations. Work at the Sian Ka'an Biosphere Reserve in Mexico includes support for a spiny lobster management project, a palm utilization study, and a nature tourism program. On the southeast coast of St. Lucia funding has supported several small-scale development projects including sea urchin egg production, firewood plantations, a fishing cooperative, and wildlife reserve management. Other model projects are under way at Gandoca Wildlife Refuge in Costa Rica and Pacayo-Samiria National Reserve in Peru.

The establishment of the Beni Biological Station and Biosphere Reserve in Bolivia is another success story in which ecologists played a role. Dr. Javier Castroviejo, a Spanish biologist, was so impressed by the diversity of the Beni area that he convinced the Bolivian government to create the 132,000-hectare Beni Biological Reserve as a natural protected area in 1982. Spanish scientists began doing research in the reserve in 1983, and in 1985 the Nature Conservancy purchased a house for the biological station. In 1986, UNESCO declared the area a biosphere reserve. In 1987, a 3.7-million-acre buffer zone was added when the Bolivian government entered into a unique debt swap with the assistance of Conservation International. A management plan is currently being developed for the reserve and buffer area by a team that includes biologists, geologists, and ecologists. The plan will address the economic needs and use of biotic resources by the Chimane Indians that live in the reserve.

*   *   *

Ecological analyses are essential to the development of appropriate biological resources-conservation projects and policies. The ecologist must combine scientific knowledge with an appreciation for the non-scientific, political considerations of international development. As is frequently noted, ecologists must improve their ability to communicate with policy-oriented disciplines and be willing to give advice on the basis of existing knowledge.

While ecological analyses contribute in many ways to our understanding of biological diversity and to the design of sustainable development projects, we still lack basic information on levels of biological diversity and on the ecological principles that sustain diversity. We know that biological diversity is the basis for essential evolutionary and ecosystem processes. Some hold that it is also a critical measure of the degree of environmental disturbance in an area and is an indicator of an ecosystem's ability to recover from disturbance.

Habitat is being converted at such a fast rate that we do not have the luxury of waiting until we have more information on the relationship between diversity and ecosystem stability. Ecologists and development professionals (one hopes that more individuals will join the ranks of both groups) should do two things immediately: begin applying what we do know to conservation and development projects; and continue demonstrating the links between conservation of biological diversity and sustainable development.

References Cited and Sources of Further Information

Agency for International Development (AID), 1987. A report on progress in conserving biological diversity in developing countries FY 1986. A report from the Interagency Task Force on Biological Diversity. Presented to the Congress of the United States, February.

AID and National Oceanic and Atmospheric Administration (NOAA), 1987. Caribbean marine resources. Opportunities for economic development and management. Washington, DC: U.S. Department of Commerce, 91 pp.

Altieri, M.A., M.K. Anderson, and L.C. Merrick, 1987. Peasant agriculture and the conservation of crop and wild plant resources. Conservation biol. 1(1):49–58.

Belovsky, G.E., 1987. Extinction models and mammalian persistence. *In:* Soulé, M.E. (ed.), Viable populations for conservation. Cambridge: Cambridge University Press, pp. 35–57.

Brady, N.C., 1986. International development and the protection of biological diversity. Speech to the National Forum on BioDiversity. Sponsored by the National Academy of Sciences and the Smithsonian Institution. Washington DC, September 24, 1986.

Council for Environmental Quality (CEQ) and the U.S. Department of State (USDOS), 1980. Global 2000 report to the President. Entering the twenty-first century (Gerald O. Barney, Study Director). Washington, DC: Government Printing Office, 3 vols.

Harrison, J., 1985. Status and trends of natural ecosystems worldwide, contract paper #6 to the Office of Technology Assessment. U.S. Congress, December. 57 pp.

International Union for Conservation of Nature and Natural Resources (IUCN), 1980. World conservation strategy: living resource conservation for sustainable development. Gland, Switzerland: IUCN.

Janzen, D.H., 1986. Guanacaste National Park: tropical ecological and cultural restoration. San José, Costa Rica: Editorial Universidad Estatal a Distancia, 103 pp.

Johnson, N., 1986. Contemporary issues in biological diversity and its conservation: a review of recent publications. International Institute for Environment and Development, draft ms. 44 pp.

Kimmerer, W.J., 1984. Diversity/stability: a criticism. Ecology 65(6):1936–1938.

Lovejoy, T.E., 1984. Application of ecological theory to conservation planning. *In:* F. di Castri, F.W.G. Baker, and M. Hadley (eds.), Ecology in practice. Paris: UNESCO, pp. 402–413.

Lovejoy, T.E., J.M. Rankin, R.O. Bierregaard, Jr., K.S. Brown, Jr., L.H. Emmons, and M. van der Voort, 1984. Ecosystem decay of Amazon forest remnants. *In:* M.H. Nitecki (ed.), Extinctions. Chicago: University of Chicago Press, pp. 295–325.

Lovejoy, T.E., R.O. Bierregaard, Jr., A.B. Rylands, J.R. Malcolm, C.E. Quintela, L.H. Harper, K.S. Brown, Jr., A.H. Powell, G.V.N. Powell, H.O.R. Schubart, and M. Hays, 1986. Edge and other effects of isolation on Amazon forest fragments. *In:* M. Soulé (ed.), Conservation biology. Sunderland, MA: Sinauer Assoc., pp. 257–285.

Macarthur, R.H., and E.O. Wilson, 1967. The theory of island biogeography. Princeton, NJ: Princeton University Press, 203 pp.

MacKinnon, J., K. MacKinnon, G. Child, and J. Thorsell, 1986. Managing protected areas in the tropics. Gland, Switzerland: International Union for Conservation of Nature and Natural Resources, 295 pp.

Mishra, H.R., and E. Dinerstein, 1987. New zip codes for resident rhinos in Nepal. Smithsonian 18(6):66–73.

National Research Council (NRC), 1982. Ecological aspects of development in the humid tropics. Washington, DC: National Academy Press, 297 pp.

Office of Technology Assessment (OTA), 1987. Technologies to maintain biological diversity. Washington, DC: Government Printing Office, 336 pp.

Oldfield, M.L., 1984. The value of conserving genetic resources. Washington, DC: Government Printing Office, 360 pp.

Orians, G.H., 1986. The place of science in environmental problem solving. Environment 28 (9):12–41.

Pielou, E.C., 1979. Biogeography. New York: Wiley, 351 pp.

Prescott-Allen, R., and C. Prescott-Allen, 1982. What's wildlife worth?: economic contributions of wild plants and animals to developing countries. London: IIED, 92 pp.

Raven, P.H., 1987. We're killing our world. Keynote address. American Association for the Advancement of Science, Chicago, February 14.

Shaffer, M.L., 1981. Minimum population sizes for species conservation. Bioscience 31 (2):131–134.

Shaffer, M.L., 1985. Assessment of application of species viability theories. Contract Paper #5 to Office of Technology Assessment, U.S. Congress. October, 45 pp.

Shaffer, M.L., 1987. Minimum viable populations: coping with uncertainty. *In:* Soulé, M.E. (ed.), Viable populations for conservation. Cambridge: Cambridge University Press, pp. 69–86.

Shmida, A., and M.V. Wilson, 1985. Biological determinants of species diversity. J. Biogeography 12:1–20.

Skorupa, J.P., and J.M. Kasenene, 1984. Tropical forest management: can rates of natural treefalls help guide us? Oryx 18(2):96–101.

Soulé, M.E. (ed.), 1987. Viable populations for conservation. Cambridge: Cambridge University Press, 189 pp.

Wilson, E.O., 1985. The biological diversity crisis. BioScience 35(11): 700–706.

Wolf, E.C., 1987. On the brink of extinction: conserving the diversity of life. Worldwatch paper 78 (June), 54 pp.

World Bank, 1986. Wildlands: their protection and management in economic development. Washington, DC: World Bank Office of Environmental and Scientific Affairs, 103 pp.

Chapter 8

THE IMPORTANCE OF APPLIED ECOLOGY IN NATIONAL CONSERVATION STRATEGIES IN THE TROPICS

JOSE I. DOS R. FURTADO

The World Conservation Strategy launched by the International Union for the Conservation of Nature and Natural Resources (IUCN) in 1980 marked a critical transition in the conservation movement. It marked a change from the fragmented conservation of nature at the local and national levels—which was isolated somewhat from the realities of human development and natural resources management—to the integration of nature conservation with human development and natural resources management at the local, national, and global levels. This change in the perception of conservation as equivalent to sustainable development, and as concerning the health of the biosphere and its component parts, may be attributable to a number of factors, notably:

• The limited success of two decades of international development in the tropical and arid zones, under the United Nations Development Decades, which emphasizes sectoral economic growth and industrialization based on models used in the temperate zone.

161

- The failure of development assistance that emphasizes economic and industrial growth in sectors to adequately address the issues of wealth distribution, environmental protection, and quality of life in the developing countries; this results in increased malnutrition, poverty, environmental hazards, graft, and debt.

- The limited success of technology transfer in the tropical and arid zones (the South) from the advanced countries of the temperate zone (the North); this involves the inappropriateness of some of the technology transferred and of the inadequacy of local scientific and technological capability for development, and therefore results in a widening technological gap.

- The discovery of ecological interlinkages between natural communities, habitats, and ecosystem or biome types at the regional and global levels, as a consequence of extensive and intensive transformation of nature and natural resources.

- Reinforcement of the notions of a common global heritage, human destiny, and cosmic unity as a consequence of advances in space science and technology, and in spiritual and psychosocial understanding of man.

- Successes in some aspects of ecological research, development cooperation, and international governance at the regional and global levels, over three decades.

The *World Conservation Strategy* (IUCN, 1980) provides essential guidelines for nature conservation and natural resources development or management in as sustainable a manner as is possible over the long term. It provides the ingredients of nature conservation that would make possible the sustainable development of natural resources, especially the protection of biological diversity and genetic resources, the sustained use of species and ecosystems, and the maintenance of essential ecological processes and life-support systems.

In this respect, the World Conservation Strategy suggests a range of actions at the local, national, and international levels; and provides a number of considerations for decision-makers who subscribe to its philosophy. However, it is not a blueprint for resolving conflicts pertaining to nature conservation, natural resources management, and human development. Rather, it is a strategy for the discovery of a variety of tactics and actions that harmonize nature conservation, natural resources management, and human development at the local, national, and international levels; it depends on psychosocial perceptions about life-style and the quality of life, and on cultural and economic factors.

NATIONAL CONSERVATION STRATEGIES

The *World Conservation Strategy* (IUCN, 1980) recommends *inter alia* that each country review the extent to which it is achieving nature conservation and natural resources management, concentrating on priority requirements for and main obstacles to it. Such a review should form the basis of a strategy for the sustainable development of natural resources. Such national conservation strategies are intended to provide the means to focus and coordinate the efforts of government agencies and nongovernment organizations in implementing the World Conservation Strategy; and to facilitate focus on priority requirements for nature conservation, to stimulate appropriate action, and to promote public consciousness to overcome any obstacles. However, there are no guidelines as to who should lead the formulation of such strategies, what steps should be taken in this process, how different partners should be involved, what should be the final outcome, and so on. This is because national conservation strategies must be related closely to the structure and culture of decision-making and policy-making in each country, and must respond to each country's level of environmental, technological, social, economic, and political awareness and capability.

Following the launch of the World Conservation Strategy, IUCN and the World Wide Fund for Nature supported its implementation particularly through the formulation of national conservation strategies, especially in the developing countries. IUCN established the Conservation for Development Center to facilitate this implementation in collaboration with its commissions, notably the Commission on Environmental Planning. The center has served as a focal point for the allocation of bilateral and multilateral aid funds for the implementation of national conservation strategies. More than 30 countries have formulated or are formulating national (or subnational) conservation strategies. Most of these are in the African and Asia-Pacific regions, are located in the tropical and arid zones, and are low-income countries. This slow pace in formulating national conservation strategies may be attributable to a number of factors, notably:

• political priority being given to short-term crises development needs (e.g., debt management, trade deficits) than to medium- and long-term conservation and development needs;

• inadequate national capability or infrastructure for assessing potential policy conflicts and complex trade-offs between nature conservation or environmental protection and natural resources development and management;
• scarcity of financial resources for nature conservation and environmental protection; and
• inadequate basic information for management and decision-making on trade-offs and conflict resolution pertaining to nature conservation, environmental protection, natural resources management, and aspirations for human development and life-style.

Nevertheless, national conservation strategies have provided a useful tool for initiating and promoting a dialogue and open-ended interaction between nature conservationists and natural resources developers and managers, about the following in particular:

• approaches for the optimal use of natural resources on a sustainable basis;
• the nature and dynamics of social and technological change;
• trade-offs (psychosocial, material, and monetary) between total exploitation of natural resources and their total preservation;
• the need for benchmark indicators or products of stages in the integration of conservation with development;
• the role of government and nongovernmental organizations, especially in the agricultural, industrial, and commercial sectors;
• intersectoral participation in the formulation, implementation, monitoring, and review stages;
• difficulties in consensus-building; and
• the need for iterative approaches for achieving a balance between development planning and socioeconomic objectives at various spatial scales (local, provincial, national, regional, and global) and different levels of technocultural impacts (localized or widespread, low or high technology).

Although the experience of national conservation strategies has been reviewed separately by Halle and Furtado (in press), this chapter discusses the importance of applied ecology in national conservation strategies in the tropics.

IMPORTANCE OF APPLIED ECOLOGY

In the formulation and implementation of national conservation strategies, applied ecology is important at three congruent levels of consideration: the general environment, the management environment, and the decision-making environment. At the level of the general environment, basic scientific information is needed about the structure and function of natural and man-made (or synthetic) ecosystems, and about cultural ecology including the transfer of science. For example: What are the geological structures, soils, water systems, climate, plants and animals, and their interrelationships and dynamics, that characterize the general environment that are being transformed, developed, and protected? What are the beliefs, values, concepts, challenges, tools, and organizations about man's relationship with himself, nature, and the universe that stress the tension field between human perception and realization, and that advance human development and nature conservation? Basic knowledge or science along these lines determines the level of human confidence, security, and fellowship, and influences the management of the environment in relation to species, ecosystems, and ecological processes. Baseline knowledge about the natural and cultural environment in the tropics is generally weak, because of meager investment in social and environmental research.

For managing nature conservation and natural resources, which is a subset of the general environment, critical information is needed about available techniques, local skills and capability, range and sophistication of equipment and supplies, local and national organizations and institutions, and about financial resources and development objectives. Information of this kind facilitates the use and protection of species, ecosystems, and ecological processes; however, such information is scarce in the tropics because of the paucity of management-oriented research.

Finally, for decision-making about nature conservation and natural resources development, which is a subset of the management environment, strategic information is required about the impacts of alternative technologies available for management on a sustainable basis, and about trade-offs and valuations, and choices and outcomes, pertaining to the enhancement of the quality of life and life-style and to resolving potential conflicts between conservation and development. Information is weakest for decision-making; and even if scientific and technological information were available, information about human values and cultural dispositions, attended by uncertainty and risks, would usually be difficult to obtain at the local and natural levels. As a consequence of the dearth of information available for proper decision-making pertaining to the management of the tropics, the instruments for nature conservation and natural resources management in this area tend to mimic those developed in the temperate zones. There is a considerable scope for human ecological and behavioral research on innovative approaches to nature conservation and natural resources management, especially those that build on human confidence, security, and fellowship at the local and national levels.

PROTECTION OF BIOLOGICAL DIVERSITY AND GENETIC RESOURCES

A large proportion of biological diversity is located in the tropics, where climatic conditions are relatively stable, where biogeophysical systems have a high metabolic rate due to high fluxes of energy (insolation) and materials transport (e.g., precipitation and circulation), and where biogeographic history has been complex.

However, only a small fraction of this diversity has been surveyed and authenticated over the past 200 to 300 years, owing to the concentration of research and development over this period in the temperate zones. Much of this authentication is restricted to dominant or spectacular plants and animals. For sustainable development of biological resources, it is necessary to survey and authenticate this rich biological diversity in the tropics, particularly in relation to taxa that are underexploited, are used traditionally in various ways and in subsistence economies, and are potentially valuable for supporting life in extreme environmental conditions. While general surveys of biological diversity may not be feasible for most tropical countries, because of prevailing economic difficulties and the scarcity of local capability, specific surveys that address taxa of potential economic and livelihood importance (e.g., fruit, timber, medicinal and aromatic plants) may be practicable. For developing countries, this may involve regional cooperation and potential end-users.

Survey and authentication of biological diversity and genetic resources are laborious and expensive, even if a utilitarian approach was to be employed. While new techniques (e.g., laser-scanning imagery, computer-aided analysis) may be needed to expedite authentication, location of centers of biological diversity or species differentiation is essential for their protection *in situ*. Such centers are likely to differ for terrestrial (and freshwater) and marine species in the tropics, because of different ecological factors that determine genetic drift and isolation on land and in the seas. Ecological techniques would facilitate the location of such centers, especially if used selectively for key taxa of herbivores, detritivores, predators, and parasites, which may serve as indicators. Although the IUCN has developed protected area systems for the use of natural resources that are well documented in national conservation strategies, such systems have not been developed for the protection of biological diversity. In fact, much of the diversity of plants and animals is outside the protected areas system, which suggests that *ex situ* protection of biological diversity may be more important than *in situ* protection.

Protection of biological diversity, particularly *in situ*, depends on an understanding of biotic potential, population dynamics, growth and life-history and the processes that maintain it, especially for key species in the dynamics of tropical ecosystems. Because of high energy and material fluxes through the grazing and decomposition pathways, and of relative environmental stability, taxa in tropical forests and seas show a high degree of specialization, of symbiotic relationships (where two species live together intimately to the benefit of both), and of life-cycle complexity—all of which make most of them highly susceptible to extinction under prolonged moderate stresses. The population dynamics of most tropical taxa are poorly understood. Basic information is available for some taxa that are exploited from natural ecosystems (for game, timber, fish, food, or trade in exotic species); and for threatened spectacular species, besides domesticated and pest species. The population dynamics of key species would enable the development of appropriate technology for *in situ* protection of biological diversity in the tropics, and the assessment of *ex situ* approaches for protection. In view of the paucity of information, this aspect of applied ecology has not been well treated in national conservation strategies.

SUSTAINED USE OF SPECIES AND ECOSYSTEMS

National conservation strategies in the tropics have examined somewhat the sustained use of species and ecosystems. The biophysical dynamics of species and ecosystems have received greater attention than the technological and management dynamics that ensure the sustainability of their use. Furthermore, the issues of sustained use have been examined to a greater degree in species harvested from natural ecosystems and in natural ecosystems, than in domesticated species and man-made or synthetic ecosystems. The population dynamics of most key species in the tropics, including those harvested commercially from the wild, is not known adequately for the development of reliable models or strategies for exploitation. Such strategies appear to be better known for animals than for plants. Some of the strategies for optimizing harvests include the maximum sustainable yield and multispecies stock assessment. In husbanded species, there is a replenishment of genetic material and variability every 7 to 15 years. We need basic information on population dynamics and genetics to facilitate optimum and sustainable production, and the assessment of trade-offs between wild-stock harvest and cultured-stock harvest. The latter enables the diversification of products and markets, including quality control; however, it requires a high level of technological and management sophistication.

Tropical ecosystems are poorly understood and few of them, if any, have been differentiated on the basis of community ecology. Ecosystem types have been differentiated generally on the basis of physical factors and dominant taxa, especially plants. Weaknesses in taxonomy and the dispersion or low density of most species in the tropics, have made detailed characterization of ecosystems difficult. Hence, assessment of their sustained use is also difficult. Protected areas have been designated for a variety of uses. Most of these are small and are not sustainable in their present form, due to edge effects. Most protected areas also experience harsh physical conditions (e.g., steep slopes) or are economically low in productivity; and there appears no congruence between their distribution and that of biological diversity or essential ecological processes, except for watershed reserves. Furthermore, the most productive ecosystems in the tropics have usually been degraded by extensive or intensive uses.

The sustained use of man-made or synthetic ecosystems is somewhat uncertain since these tend to become increasingly disorganized by human perturbations. Therefore, we must use a biogeographic approach when designing and distributing protected areas, especially in relation to the protection of biological diversity *in situ* and the maintenance of essential ecological processes.

MAINTENANCE OF ESSENTIAL ECOLOGICAL PROCESSES

National conservation strategies in the tropics have dealt poorly with ecological processes essential for supporting life, since these are poorly understood. There are two kinds of processes: the essentially abiotic and the biotic. The abiotic processes include the carbon-dioxide/oxygen cycles and the hydrological and nutrient cycles. They depend on atmospheric, oceanic, and physiographic circulation, in which animals and plants are intimately involved. They are better known than the biotic processes essential for supporting life. The latter include production in relation to grazing, decomposition in relation to detritivory, predation and parasitism in relation to population control and fitness, thermoregulation in relation to habitat structure and stratification, and a variety of symbiotic relationships involved in the alternation of generations, growth and maintenance, reproduction including pollination, and the formation of secondary metabolites. These ecological processes are better known in terrestrial than in marine systems. They are interrupted by human technologies or synthetic ecosystems. They can be maintained by using approaches in landscape or seascape ecology to integrate resource-use units with natural units and to generate a landscape or seascape mosaic that is complex, stratified, and mimics natural ecosystems.

Tropical ecosystems are the most complex, most fragile, and least understood ecosystems for use for human development, primarily because of the concentration of research effort in the temperate and boreal zones over the past 200 to 300 years. There is a critical need for the application of population and community ecology, landscape and systems ecology, and human ecology pertaining to management and decision-making, to ensure the sustainability of nature conservation and natural resources development. This application is needed not only in the formulation of national conservation strategies but in their implementation and in development planning.

References Cited and Sources of Further Information

Furtado, J.I., 1979. The status and future of the tropical moist forests in South East Asia. *In:* C. MacAndrews and L.S. Chia (eds.), Developing economics and the environment: the southeast Asian experience. Singapore: McGraw-Hill International, pp. 73–120.

Furtado, J.I., and K. Ruddle, 1986. The future of tropical forests. *In:* N. Polunin (ed.), Ecosystem theory and application. New York: Wiley, pp. 145–171.

Golley, F.B., J.T. McGinnis, R.G. Clements, G.I. Child, and M.J. Duever, 1975. Mineral cycling in a tropical moist forest ecosystem. Athens, GA: University of Georgia Press, 272 pp.

Goodland, R.J.A., and H.S. Irwin, 1975. The Amazon jungle:green hell to red desert? Amsterdam: Elsevier Scientific Publishing, 155 pp.

Halle, M., and J.I. Furtado, in press. The role of national conservation strategies in attaining the objectives of the World Conservation Strategy. Proceedings of the International Conference on the World Conservation Strategy, June 1986, Ottawa.

Holdridge, L.R., and J.A. Tosi, 1972. The world life-zone classification system and forestry. San Jose: Tropical Science Centre, 24 pp.

IUCN (International Union for Conservation of Nature and Natural Resources), 1980. World conservation strategy, Gland, Switzerland: IUCN, 54 pp.

Janzen, D., 1973. Tropical agroecosystems. Science 182(4118):1212–1219.

Janzen, D., 1986. The future of tropical ecology. Ann. Rev. Ecol. Syst. 17:305–324.

N.R.C., 1980. Report of survey of conversion in tropical moist forests. Washington, DC: National Research Council, 205 pp.

Nye, P.H., and D.J. Greenland, 1960. The soil under shifting cultivation. Harpenden, England: Commonwealth Agricultural Bureau, 156 pp.

Reichle, D.E., J.F. Franklin, and D.W. Goodall (eds.), 1975. Productivity of world ecosystems. Washington, DC: National Academy of Sciences, 166 pp.

Chapter 9

ENVIRONMENTAL SUSTAINABILITY IN ECONOMIC DEVELOPMENT —WITH EMPHASIS ON AMAZONIA

ROBERT GOODLAND

Although the linkages between economics and environment are absolute, these linkages are not emphasized either by the economic or by the environmental professions. The economics of natural resources remains a minor unpopular theme in orthodox economics today. Similarly, the relationship of the economic subsystem of the overall ecosystem is only mentioned in most environmental science texts. Attention to the environmental dimensions of economic development began in the early 1970s and has yet to become systematic. There are encouraging signs that this is improving. The integration of economics and environmental concerns has been the focus of several recent conferences and books and at least two journals (*Environmental Economics,* 1976; and *Ecology and Economics,* 1987); excluding the vast literature on mainly reversible and pollution economics. This chapter discusses four environmental aspects

171

of economic development, the concept of sustainability, carrying capacity, ethics, and irreversibility, and one important irreversible—the loss of biodiversity—and relevant World Bank's policies.

SUSTAINABILITY

Sustainable development may be defined as a pattern of social and structural economic transformations (i.e., development) that optimizes the economic and other societal benefits available in the present, without jeopardizing the likely potential for similar benefits in the future. A primary goal of sustainable development is to achieve a reasonable and equitably distributed level of economic well-being that can be perpetuated continually for many human generations. Sustainability implies a transition away from economic growth based on depletion of nonrenewable resource stock (e.g., oil) and toward progress (i.e., improvement in the quality of life) based more on renewable resources over the term.

This definition of sustainability has five major implications:

1. *Well-being* depends on three categories of values, which development should consider, although some trade-offs are inevitable: economic efficiency; equitable distribution of economic resources; and noneconomic values, such as religion and spirituality, dignity, esteem, esthetics, and liberty.

2. *Intergenerational equity.* Although it is impossible to predict with precision the likely interests of future generations, it is prudent to assume their need for natural resources (soil, air, water, forests, fisheries, biodiversity, energy, and minerals) will not be markedly less than ours. Therefore, sustainable development implies using renewable resources in a manner that does not degrade them—harvesting renewables on a sustained-yield basis rather than mining to near extinction. Whales, tropical forests, and coral reefs are examples of renewables that are often mined rather than harvested sustainably. High discount rates (commonly 10 percent and more) discourage long-term investments and promote short-term investments. For example, if a whaler can make a 15 percent per annum profit by exterminating whales over 10 years, and then investing the proceeds in a different sector, what economic incentive is there to make a 10 percent per annum profit by harvesting whales sustainably?

3. *Sustainable development* implies using nonrenewable (exhaustible) mineral reserves in a manner that does not unnecessarily preclude easy access to them by future generations. It will be easier in the future to use today's scrap metal if it is recycled, rather than dumped in a dispersed manner.

4. *Sustainable development* also implies depleting nonrenewable energy resources at a slow enough rate to ensure a high probability of an orderly societal transition to renewable energy sources (including solar, biomass, wind, and hydroelectric) when nonrenewable energy becomes substantially more costly.

5. *Agricultural sustainability* implies the permanent maintenance of biological productivity on the site, with the costs of imported inputs (e.g., diesel, biocides) and nutrients (e.g., fertilizer) not exceeding the commercial value of the site's production. Sustainability demands resilience, the ability to return to an initial state; resistance or ability to cope with change (such as the addition or reduction of species or weather cycles); and persistence, which means that the agroecosystem succeeds for a long time relative to the life spans of its components.

These sustainability factors are not new to most people. The more concerned and thoughtful individuals already drive smaller cars, conserve energy, recycle waste, provide for our descendants, and behave thriftily. As a society, though, we allow ourselves and our globe to be operated on a system of frontier economics created when the world was essentially empty of people, when resources were scarcely tapped, air and water were clean, land and forests were abundant. This idyllic era has long since vanished, but the economic orthodoxy remains to be modernized to reflect today's world, which is full (or overfull). Possibly the single most influential change toward sustainability is to revamp neoclassical economics to accommodate environmental concerns, particularly ethical and nonhuman values, sustainability, and externalities.

ENVIRONMENT AND ECONOMICS

In most introductory college economics texts, economics is still viewed as an isolated circular flow of production and consumption. (The scope for capital and labor to substitute for resource inputs is clearly limited.) This is like an ecologist's describing an animal's circulatory system in detail, while omitting to note the digestive tract and the respiratory system. Economists have to realize that the economic circulation of production and consumption also has a digestive tract; that this is one-way, and that it is irreversible. The economic subsystem takes in resources and excretes wastes, hence is irrevocably and tightly linked to the ecosystem. The economic subsystem depends on the ecosystem as a source of raw materials and as a sink for wastes: both these are finite.

Therefore the main variable is the one-way throughput of matter and energy; this throughput originates in the ecosystem sources, is used in the economic subsystem, and is expelled back into the ecosystem's sink. Addressing the dependence of the economic subsystem within the ecosystem raises the question of how big the economy should be in physical dimensions relative to the ecosystem. Now that the source and sink functions of the ecosystem are overstressed, the scale of the throughput has started to matter. This questions the concept of growth economics, gross national product, and its relation to welfare, equity, and the impossibility of generalizing Western standards to the entire world. In the United States alone 6 percent of the world's people use 30 percent of the world's nonrenewable resource flows. Clearly it would be impossible for the other 94 percent of the world's population to spend resources on the same scale.

Therefore we need to define economics and economic development in a way that applies to everyone: this leads us back to throughput. Economic development must be redefined less in terms of expanding GNP; more in terms of controlling population; and certainly in terms of limiting inequality.

CARRYING CAPACITY

Large parts of our planet already have exceeded their carrying capacity. Humanity can support itself (often none too well, either, since many thousands starve to death every day) only by consuming its inherited capital. Not only are stocks of oil and minerals being rapidly depleted, but topsoils, groundwater, and biodiversity are also declining. Those who care for the earth must grapple with hard trade-offs about humans. What will be the quality of life of the 8 billion to 16 billion people projected for the next century?

In June 1987 alone, if 3,186 tons of New York garbage was barged 6,000 miles for Long Island, refused by six states and three nations (Mexico, Belize, and the Bahamas) for over four months, and still no backyard was found in which to dump it, there cannot be much unused space (sink) left. Since then it has been incinerated, but even the ash may be too toxic to bury in a Long Island, New York, landfill. The United States produces 150 million tons of trash each year, and 90 percent of that ends up buried.

Also in June 1987, the U.S. Nuclear Regulatory Commission failed to rent a long-term radioactivity dump in spite of multimillion-dollar annual baits to several southern and western townships. The state of Nevada refused a federal offer of 100 million dollars annually for 50 years to accept spent nuclear fuel. Such high-level radioactive waste is accumulating from more than 100 power plants at 3,000 tons per year and remains dangerous for 10,000 years. Brundtland's (1987) conclusion is the sanest. Develop nuclear power only when the problems are solved beforehand: prudent advice for all products. Until society progresses to that stage, we should halt making any more unmanageable waste and force the producers (utilities) to solve their own problems. Should businesses incapable of coping with their own products be tolerated? Since 1988 marks one decade without a single reactor being ordered, possibly our society is indeed improving.

You can read this chapter in about 60 minutes, which at 150 human births per minute, means 9,000 babies were born since you began. The main problem with us humans is that we are doing too well. Yes, poverty and environmental damage are related. But the latter is more fundamental, and the affluent arguably opt to cause more environmental abuse than the poor are forced to do.

The search for unfilled sinks may reduce ocean dumping and may buy time to improve technology so that it generates less waste. For example, West Germany's largest nuclear industry, Kraftwerk-Union, signed an agreement with China in July 1987 to bury nuclear waste in Inner Mongolia's Gobi Desert in exchange for nuclear technology. Similarly, Amalgamated Shipping's *Khian Sea* (Bahamas) has been sailing for 20 months since October 1986 loaded with 15,000 tons of what the Guinean government calls toxic waste from Philadelphia. Norway's consul general and several Guinean Ministry of Commerce officials have been jailed for complicity in dumping wastes in Guinea. They were formerly rejected by the Bahamas, the Dominican Republic, Honduras, Haiti, and Guinea-Bissau. This sounds unsatisfactory unless (and originally even if) it is a brief stop-gap expedient. Otherwise, there is little incentive to recycle (or to operate without generating undisposable wastes) until all of the world's deserts are overflowing. The world can never be a decent place to live in until there is respect, tolerance, and appreciation for everything on it.

Much of the world has overshot its carrying capacity. Until life becomes more livable in the tropics, people will abuse them. This is why people who care for the environment strenuously promote social justice in the tropics. If demographic inertia, the greenhouse effect, acid rain, and loss of tropical forests are included, then this claim becomes ominously plausible. This is the case for a shift from growth economics to steady-state or sustainable economics; a shift from quantity to quality.

Much of the joy—and the grief—of ecology is that most environmental stresses are fundamentally different from conventional cause-and-effect expectations. Overshoot is a case in point. The environment appears to have many thresholds of which stresses appear linear. Once the threshold is exceeded, the effect may be nonlinear, or irreversible, or lagged, or cumulative, or compounding, or even synergistic. In the case of Europe's Waldesterben: Why did forests tolerate decades or centuries of pollution with little sign of stress, only to succumb in vast tracts in the space of a couple of years? The basis for decisions should be: "How bad would it be if you were wrong?" What will it take for us to learn to err on the side of prudence in the face of so much uncertainty and to choose a course of action likely to do the least damage if our understanding proves faulty?

ETHICS

"We didn't inherit this earth from our parents; we're borrowing it from our children. . . ."
                                                    —attributed to DAVID BROWER

Ethics pervade both economic development and environment, although we do not always acknowledge the ethical component of decisions. The relationship among growth, equality, and ethics is fundamental to environmental management in economic development. Just as it is impossible for everyone's income to be above average, so is it impossible for everyone's relative income to increase as a result of growth. In rich countries, well-being increases with relative income, not with absolute income. Therefore, aggregate growth above sufficiency leads to self-canceling effects on well-being.

If aggregate growth does not improve well-being, then it is undesirable, even if it may be possible. Aggregate growth is undesirable because it augments throughput in the economic subsystem, hence strains the sources and sinks of the overall ecosystem. It strains sources in general, and common property sources in particular, since markets are imperfect. Throughput, or the "scale decision," is a social decision. The rate of takeover of habitats and the rate of drawdown of geological capital are decisions concerning sustainability, intergenerational equity, or justice; hence, the major need to integrate ethical criteria into economics and into societal decision-making in general. Although related, sustainability and intergenerational equity, may need two scales of value.

In 1986, World Bank vice-president Shahid Husain highlighted the ethical component of economic development: "The discounting procedures we use to make judgments how fast we can, or should, deplete natural resources are basically moral value judgments. Explicitly or implicitly, we make a moral judgment when we say that this generation, or any generation, has the right to make decisions for the future simply by having an adequate discount rate."

Although much of this talk is too anthropocentric, the similarly important ethical and intrinsic values of nonhuman species must be acknowledged. Be under no illusion that incorporation of the ethical dimension into development will be easy. John Shad, chairman of the Securities and Exchange Commission, offered Harvard University Business School 20 million U.S. dollars, as a gift to introduce a course of business ethics. Harvard rejected the gift. Ethics is both agonizing and enthralling. The basic point is to embrace diversity and tolerate logical extremes. I confess my uneasiness with those cool logicians who point out the inescapable need to eschew all pets (dogs, cats, goldfish, budgerigars) and all domestic animals (cows, sheep, pigs, horses, chickens). Singer (1975, 1985) and Regan (1987) prove to me that the intrinsic values of nonhuman species preclude their exclusively utilitarian use. We accept the rich environmental ethics spectrum or diversity of opinion ranging from one logical extreme (e.g., Singer and Regan); to vegans and vegetarians; to people who eschew terrestrial carnivory, leather shodding, and fur cloaking; to people who effectively conserve many habitats and raise bundles of environmental financing solely for the right to cull a few species over a few days a year (e.g., National Wildlife Federation, Ducks Unlimited, Izaak Walton League). Our ethical colleagues have to analyze where they are comfortable. Some wear fur only if domestically bred rather than harvested from the wild. Gandhi made, presented, and wore leather sandals, but only from cows dying from old age. He even condoned mosquito killing for malaria control because it was like a fence animals need not cross.

Wherever on this spectrum you fit, it won't be for long. Ethical values evolve. I believe the global society gradually and too slowly is improving ethically: patchily and with reverses. Flogging your horse to death, human slavery, spouse and child bashing were legal until relatively recently. Nowadays starvation, absolute poverty, extinction of species, sadistic blood sports (e.g., cockfighting in southern United States; fox hunting in the United Kingdom), and street people freezing to death, while lamentably legal, at least needle the conscience of society more than they did a century ago. Some of our societal institutions seek to redress such evils. Other ethical grotesqueries remain less acknowledged; for example, the heinous imbalance between education, nutrition, social security, health, potable water, etc., and military expenditures on the other. Just as the recently acknowledged environmental dimension of economics needs attention, so do the ethical components of both economic and environmental concerns. The global commons is run along increasingly economic criteria. The time is overdue to complement this with environmental and ethical considerations. The tremendously encouraging consonance between the two occurs because the ethical and environmental solutions in economics are almost always similar.

Clearly, population issues are fundamental to sustainability and to environmental management. I shall not add to the vast literature on the subject, but this chapter would be incomplete without mention of two fundamentals: the liberation of women and the excess consumption of the affluent.

The liberation of women is essential to environmental sustainability, as well as being a key equity issue. The president of the World Bank recently emphasized that "women do two-thirds of the world's work. Their work produces 60 to 80 percent of Africa's food, 40 percent of Latin America's. Yet, they earn only one-tenth of the world's income and own less than 1 percent of the world's property." Hence, women constitute an unacceptably large proportion of the poorest of the poor, and of the landless laborers, in addition to the burden of the double day: both job and housekeeping. Resolution of this impasse lies in education and in improvements in agriculture—particularly reuniting nutrition with agriculture—and in agrarian reform. Even modest increase in self-reliance and self-sufficiency buffer the family and augment equity. The Bhopal tragedy, for example, represents agribusiness profit motives more than human subsistence. Bearing fewer children must be made a reasonable choice for women; at the moment it is not. The wish for fewer children will increase when progress toward the above has been achieved. Fostering a coalition between groups that promote environment, women, and population management would help greatly.

The second fundamental is to reduce the pressures on our finite globe by reducing excess consumption by the affluent. Intergenerational equity can be approached only when we do not deprive future generations of necessities in order to pamper ourselves with luxuries today. The gigantic volume of nonrenewable resources that deplete Third World stocks is galvanized largely through demand for luxuries. The old saying remains valid: "The United States' 6 percent of the world's population consumes 30 percent of the world's resources." This is clearly impossible to generalize for the other 94 percent of the world's population: only imagine 1 billion Chinese with one automobile each. Let me end these two points with a question: If walls were built around each nation (i.e., if all international trade were halted), which would crumble first, the industrial or the developing countries?

IRREVERSIBILITY

Environmental effects can be reversible or irreversible. With the blinding clarity of hindsight, we now see that the environmental movement began with reversible environmental effects. The best example is pollution in an industrial region, such as Ohio's Cuyahoga River, which was operated as a petrochemical sink and died decades ago. It was so polluted that it occasionally even caught fire. But shortly after the political will and financing were mustered, the pollution was reversed, the river was repopulated with wildlife indistinguishable from pre-pollution eras, and today it is attractive and clean. Similarly in London, England, the political will to combat the notorious pea-soup smog was mustered by 4,600 extra deaths in one night alone (December 4, 1952). This led to the 1956 Clean Air Act; today London's air is relatively clean. The lesson here is that damage and death of temperate ecosystems can be cured. This is a needlessly expensive course, but it is possible. The dental analogy is somewhat similar. You may hate the dentist and never seek preventive maintenance, but once the pain in your tooth exceeds a certain threshold (the political will), you voluntarily take yourself to the dental chair. Economics has contributed greatly to reversing reversibles, for example, the "polluter pays" principle and many other neat economic incentives, permits, taxes, and penalties.

Irreversible effects are fundamentally different. They cannot be cured by throwing money at them; prevention seems the only course. Examples include global carbon dioxide accumulation, the greenhouse effect, global warming, atmospheric ozone depletion, contamination of groundwater, accelerated loss of topsoil, and particularly tropical deforestation, extinctions, and loss of biodiversity. Civilization can survive the consumption of the last drop of oil reserves but not the continuing loss of topsoil. Volumes have been written on all these effects. The main cause of the extinction of species or loss of biodiversity is the loss of tropical forest, the planet's richest terrestrial ecosystem. At best, extinctions mean resources that otherwise might improve the quality of human life will not be available; at worst, it could mean serious disruption of the environmental processes upon which civilization depends.

THE LOSS OF BIODIVERSITY

> *"The poverty of the poor—and the greed of the rich—are the greatest polluters. . . ."*
>
> —INDIRA GANDHI

The loss of biodiversity or the extinction of species is the paradigm of lack of sustainability. The main cause is loss of tropical forest— dwindling at 74,000 acres per day—the planet's richest terrestrial ecosystem. This habitat supports half of the world's 5 million to 30 million species on 7 percent of our land area. Latin America alone may support 15 million species, possibly one-third of the world's total. The causes, particularly poor people forced into shifting cultivation of tropical forest, and the perilous consequences of loss of biodiversity are well known; others can turn to the Ehrlich and Ehrlich's (1981) compelling oeuvre. Lest U.S.-based readers feel it is not their problem: it is. The United States' inexorable demand for tropical commodities such as rubber, beef, oil palm, cocoa, tropical timbers (1 billion U.S. dollars per year), and veneer exerts irresistible pressure on tropical forests. If conversion of tropical forests created sustainable jobs or permanent benefits, it would be difficult to demote; but it does not. Tropical forest is being destroyed for ephemeral pittances. One of the most damaging is

The Hamburger Connection

If cattle gain 50 kilograms per hectare per year and are slaughtered after 4 years (but let's assume 8 years), and half the weight is nonmeat (skin, bones, etc.), then each cow produces 200 kilograms, or 1,600 hamburgers. If 1 hectare of tropical moist forest weighs 800 tons, then 800 tons = .5 tons of forest per hamburger. At 10,000 square meters per hectare − 1,600 = 6.25 square meters of forest per hamburger. After 10 years, this means 3 U.S. dollars per hectare per year. In summary, one hamburger translates into half a ton of forest. The cumulative effect of such hamburger consumption is equivalent to millions of years of evolution and to thousands of species. Is it worth it?

Conversion of all of Amazonia (4 million square kilometers) to cattle pasture would produce (1,600 hamburgers per hectare for 8 years) $6.4 \times 10''$ hamburgers, enough for 3 burgers per day for the world's 5 billion people for 43 days ($6.4 \times 10''/5$ billion = 128 burgers per person × 3 = 43 days). Would the irreversible loss of the Amazon forest for a month of hamburger for the world's population be a worthwhile trade-off?

Cost Item (Billions)	Cost Per Hamburger (U.S. dollars)	Total (U.S. dollars)
SUDAM tax credit programs	1.2	.135
Central bank rural credits	1.5	.083
Timber destruction	4.5	.524
742 U.S. dollars	TOTAL SOCIAL COSTS:	
	7.2 billion U.S. dollars	

For every quarter-pound unit of Amazon beef that costs Brazilian ranchers .26 U.S. dollars to produce, Brazil absorbs .742 U.S. dollars in social costs. Stated differently, 1 metric ton of Amazon beef (before processing) embodies about 6,500 U.S. dollars in social costs (of which subsidies alone represent 1,900 U.S. dollars). In stark contrast, between 1971 and 1982, Brazil paid for foreign-produced and -processed beef an average import price of 1,086 U.S. dollars per metric ton. The Brazilian government could have acquired nearly 2 metric tons of foreign beef for the same amount of capital it spent subsidizing the production of 1 metric ton of Amazon beef.

SOURCE: After Uhl, 1986; Uhl and Parker, 1986; and Browder, 1988.

deforestation for a few years of minutely cheaper beef hamburgers (see page 182). This is one of the most serious causes of forest loss in Central America.

Today's extinction rates are unprecedented; the problem is urgent. Our global society has 10 to 30 years left to conserve tropical rain-forest habitats, which are dwindling at 200,000 square kilometers per year, depending on the site. For the Philippines, Haiti, El Salvador, Sri Lanka, parts of India, and Bangladesh it is largely too late: a large proportion of their biodiversity is already irreversibly extinct. If *our* generation does not conserve samples, it won't be there for the next. In addition, most extant species may be living dead (i.e., no longer breeding or with nonviable populations). Major needs to arrest these losses are summarized in Appendix A.

ECONOMIC PROGRESS

> *"The worst thing that can happen will happen. In the 1980s this is not energy depletion, economic collapse, limited nuclear war, or conquest by a totalitarian government. As terrible as these catastrophes would be for us, they can be repaired in a few generations. The one process ongoing in the 1980s that will take millions of years to correct is the loss of genetic and species diversity by the destruction of natural habitats. This is the folly our descendants are least likely to forgive us."*
>
> —EDWARD O. WILSON, professor of biology, Harvard University

The question becomes: What are we—not our children—going to do about all this? John Maynard Keynes, as one of the founders of the World Bank, stressed: "We *will* do the rational thing—but only after exploring all other alternatives." Appendix A outlines ways to save tropical forests and hence reduce the extinction rate. We have many opportunities to become involved. A wide spectra of skills and energies are needed. Following 1987's environmental campaign, Coca-Cola's Minute Maid orange juice subsidiary donated 40,000 acres of Belizean tropical forest for conservation, plus 50,000 U.S. dollars to the Massachusetts Audubon Society toward its management. Coca-Cola now seems to be the only major multinational to conserve a sample of the natural ecosystem from which it derives profit. Can other corporations involved in tropical forest loss be similarly encouraged? The success of Earth First and the Rainforest Action Network, encouraging Burger King to eschew tropical rain forest beef, attests the possibilities. Choose where you feel most comfortable and build on your strengths.

Political actions can be exceptionally influential. Improvements in economic calculus reap overall fundamental benefits. The World Bank's President Conable wrote (1987) that "the goals of poverty alleviation and environmental protection are not only consistent, they are interdependent. Sound ecology is good economics. The Bank will place new emphasis on correcting economic policies that promote environmental abuse."

Ecologists must learn to communicate with economists and decision-makers. The World Bank's vice-president, Shahid Husain, amplified: "Many direct governmental subsidies are unsound both in environmental and in economic terms. They add to a country's fiscal burden, encourage the wasteful use of scarce resources and frequently benefit (only) the large landowners." Brazil's two major programs to subsidize beef cattle production in the Amazon, combined with the destruction of forest resources those programs encourage, represent a total social cost of no less than 7.2 billion U.S. dollars.

POLICIES OF THE WORLD BANK

The World Bank is the largest investor in economic development in developing countries. It is owned by 151 member countries; therefore practically all the world's biodiversity is contained therein. It lends 17 billion U.S. dollars per year to the developing-country sovereign owners of tropical forests. Therefore, the World Bank can be enormously potent for world environmental sustainability, by financing loans to developing nations for sustainable development, including the conservation of biodiversity.

In July 1987, the World Bank created a top-level Environment Department, in addition to four regional environmental divisions (for Africa, Asia, Latin America, and the rest). As Saul Alinsky advised: "The best way to change an institution is to force it to abide by its own rules." The World Bank's main environmental policies, which are obligatory in all World Bank–assisted programs, are listed below.

- Comprehensive environmental policy
- Involuntary resettlement of people
- Tribal people (unacculturated ethnic minorities)
- Wildlands conservation (genetic diversity)
- Cultural property (archeological and historic sites)
- Industrial safety and accidents
- Pollution control (air, water, ground, industrial, urban)
- Pesticides
- Health precautions in pesticide use
- Dams and reservoirs (hydro and irrigation projects)
- Environmental assessment

In September 1986, the World Bank promulgated a fundamental policy on biodiversity called Wildlands. So far, the World Bank has financed biodiversity preservation in 26 countries worldwide, in 40 projects, which protect more than 60,000 square kilometers (larger than all of Costa Rica). The biggest is in Amazonian Brazil, where in Rondonia protection of 19,000 square kilometers of tropical forest (as big as El Salvador) is being financed. But this is a drop in the bucket and scarcely enough to decelerate biodiversity loss. To change an institution by making it abide by its own rules, you must know the rules. Here are the specifics of the new Wildlands Policy:

The general policy follows the first injunction of Hippocrates 2,000 years ago: *Non noli nocere*—First do no harm. The World Bank seeks to avoid eliminating wildlands and wishes to finance their preservation. This translates into six policy elements:

THE FIRST POLICY ELEMENT. The Bank normally declines to finance conversion of wildlands of special concern: these are the officially designated areas such as national parks, and includes areas slated for official designation, as well as specific geographical priorities, and specific endangered ecosystems.

Why is the word "normally" included? This provides for a powerful *quid pro quo:* a trade-off. Let's say 1 percent of a desert national park is requested for an irrigation reservoir. The trade-off could include:

- Addition to the park of 1 percent equivalent tract elsewhere.
- Selection of an underrepresented habitat.

- Financing for the whole park.
- Addition of an aquatic habitat and source of water.

THE SECOND POLICY ELEMENT. When wildlands other than those of special concern become involved, the project should be sited on already converted land (logged over, degraded, or already cultivated). Rehabilitation of degraded sites must take precedence over further conversion of intact wildlands.

THE THIRD POLICY ELEMENT. All deviations from these policies must be explicitly justified. This means a more open decision-making process than has occurred in the past. The options must be aired outside the implementing ministry, and should include the Ministry of Environment, academia, and nongovernmental organizations.

THE FOURTH POLICY ELEMENT. When development of wildlands is explicitly justified, less valuable wildlands are to be converted, rather than more valuable ones. If nothing else, all these policy elements will vastly augment employment for tropical, applied ecologists and improve our knowledge of biodiversity.

THE FIFTH POLICY ELEMENT. When conversion of wildlands is not of special concern, such wildland loss must be compensated for by financing the preservation of a similar wildland elsewhere, similar in area and in biodiversity. Often this translates into creating a protected area adjacent to the project buffer zone. But it can include financing of a needed tract elsewhere if the trade-off is firmly negotiated.

THE SIXTH AND FINAL POLICY ELEMENT. If a project does not contemplate conversion of wildlands, the policy is to preserve wildlands for their environmental service values alone. The best example of an environmental service component is Dumoga National Park in Sulawesi, Indonesia. This 3,200-square-kilometer National Park—bigger than Luxembourg—protects the entire watershed of a 197-square-kilometer irrigated rice scheme, even though the rice project would not have directly caused deforestation of the watershed. The park was financed from 1.5 percent of project costs (at 1.1 million U.S. dollars). Since then, it has become the site of the world's biggest-ever scientific expedition—Operation Wallace. Hundreds of scientists worked for three years and discovered hundreds of new species.

It is up to all of us to maintain constant vigilance, to work with the ministries of environment and wildlife departments of developing countries, as well as with the implementive ministries and development finance agencies. Of course there will be difficulties, but cooperation between governments, developing-country nongovernmental organizations, development agencies, the informed public, academia, and the private sector will ensure success. Two difficulties remain: first that the World Bank makes loans at near commercial interest rates, rather than making grants, as some bilaterals fortunately can do. Second, too much loan conditionality encourages potential borrowers to seek less conditional loans elsewhere. While environmental improvement will not be easy, it is better than the alternative of accelerating extinction.

References Cited and Sources of Further Information

Akers, K., 1983. A vegetarian sourcebook. New York: Putnam, 229 pp.

Binswanger, H.P., 1987. Fiscal and legal incentives with environmental effects on the Brazilian Amazon, discussion paper ARU69. Washington, DC: World Bank, 41 pp.

Brown, B.J., et al., 1987. Global sustainability: toward definition. Environmental Mgmt. 11(6):713–719.

Brundtland, G.H., 1987. Our common future (U.N. World Commission on Environment and Development). Oxford: Oxford University Press, 383 pp.

Caufield, C., 1985. In the rainforest. New York: Knopf, 208 pp.

Conway, G.R., 1985. Agroecosystem analysis. Agric. Admin. 20:31–55.

Daly, H.E., 1985. The circular flow of exchange value and the linear throughput of matter-energy: a case of misplaced concreteness. Rev. Soc. Econ. 42 (3):279–297.

Dankelman, I., and J. Davidson, 1988. Women and environment in the Third World. London: Earthscan, 210 pp.

Davidson, J., 1985. Economic use of tropical moist forest, Commission on Ecology Paper 9. Gland, Switzerland: IUCN, 28 pp.

Dyer, J.C., 1982. Vegetarianism: an annotated bibliography. Metuchen, NJ: Scarecrow Press, 280 pp.

Ehrlich, P.R., and A.H. Ehrlich, 1981. Extinction: the causes and consequences of the disappearance of species. New York: Random House, 305 pp.

Fearnside, P.M., and G. de L. Ferreira, 1984. Roads in Rondonia: highway construction and the farce of unprotected reserves in Brazil's Amazonian forest. Environmental Conservation 11(4):297–360.

Friends of the Earth (FOE), 1987. The good wood guide (with National Association of Retail Furnishers, 17 George St., Croydon, U.K.), London: FOE.

FOE, 1987. A hard wood story: Europe's involvement in the tropical timber trade. London: FOE, 118 pp.

Foreman, D., and B. Haywood, 1987. Ecodefense: a field guide to monkey-wrenching. Tucson, AZ: Ned Ludd Books, 312 pp.

Fornos, W., 1987. Gaining people, losing ground: a blueprint for stabilizing world population. Washington, DC: Population Institute, 121 pp.

Goodland, R., H.S. Irwin, and G. Tillman, 1978. Ecological development for Amazonia. São Paulo, Ciencia e Cultura 30(3):275–289.

Goodland, R., 1988. A major new opportunity to finance biodiversity preservation. *In:* E.O. Wilson (ed.), Biodiversity. Washington, DC: Smithsonian.

Goodland, R., and G. Ledec, 1987. Neoclassical economics and principles of sustainable development. *In:* R. Costanza and H.E. Daly, (eds.), Ecological Economics. Amsterdam: Elsevier, pp. 19–46

Hardin, G., and J. Baden, (eds.), 1977. Managing the commons. San Francisco: Freeman, 294 pp.

Hargrove, E.C. (ed.), 1986. Religion and the environmental crisis. Athens: University of Georgia Press, 222 pp.

Hecht, S., 1985. Dynamics of deforestation in the Amazon. Washington, DC: World Resources Institute, 48 pp.

Hoban, T.M., and R.O. Brooks, 1987. Green justice: the environment and the courts. Boulder, CO: Westview Press, 250 pp.

Janzen, D.H., 1973. Tropical agroecosystems. Science 182(4118):1212–1219.

Janzen, D.H., 1986. The future of tropical ecology. Ann. Rev. Ecol. Syst. 17:304–324.

Jansson, A.M. (ed.), 1984. Integration of economy and ecology: an outlook for the eighties. Stockholm: Wallenberg Foundation, 240 pp.

Lal, R., 1986. Conversion of tropical rainforest: agroeconomic potential and ecological consequences. Adv. Agron. 39:173–264.

Ledec, G., 1985. The political economy of tropical deforestation. *In:* H.J. Leonard (ed.), Divesting nature's capital: the political economy of environmental abuse in the third world. New York: Holmes and Meier, pp. 179–226.

Ledec, G., and R. Goodland, 1989. Environmental perspective on tropical land settlements. *In:* D.A. Schumann and W.L. Partridge, (eds.), The human ecology of tropical land settlement. Boulder, CO: Westview Press.

Mahar, D. (ed.), 1985. Rapid population growth and human carrying capacity: two perspectives, Staff Working Paper 960. Washington, DC: World Bank, 89 pp.

Midgley, M., 1983. Animals and why they matter. Athens: University of Georgia Press, 305 pp.

Norton, B.G. (ed.), 1986. The preservation of species: the value of biological diversity. Princeton, NJ: Princeton University Press, 305 pp.

Office of Technology Assessment (OTA), 1987. Technologies to maintain biological diversity. Washington, DC: Government Printing Office, 336 pp.

Pearce, D.W., 1985. Sustainable futures: economics and the environment. London: University College, Inaugural Lecture (5 Dec), 30 pp.

Pillet, G., and T. Murote (eds.), 1987. Environmental economics: the analysis of a major interface. Geneva: R. Leingruber, pp. 1–10.

Pollock, C., 1987. Mining urban-wastes: the potential for recycling. Washington, DC: Worldwatch Institute, 58 pp.

Randall, A., 1981. Resource economics: an economic approach to natural resource and environmental policy. Columbus, Ohio: Grid Publications, 415 pp.

Redclift, M., 1984. Development and the environmental crisis. London: Methuen, 149 pp.

Regan, T., 1983. The case for animal rights. Berkeley: University of California Press, 425 pp.

Regan, T., 1987. The struggle for animal rights. Clarks Summit, PA: International Society for Animal Rights, 208 pp.

Repetto, R., 1987. Population, resources, environment: an uncertain future. Population Bull. 42(2):1–44.

Repetto, R., and M. Gillis (eds.), 1987. Public policy and the misuse of forest resources. Cambridge, MA: Cambridge University Press.

Rich, B., 1985. Multinational development banks: their role in destroying the global environment. The Ecologist 15(1, 2):21–38.

Rich, B., 1986. Multilateral development banks and the environment. Ecol. Law Q. 12(4):681–745.

Robie, D., 1986. Eyes of fire: the last voyage of the Rainbow Warrior. Philadelphia: New Society, 168 pp.

Rosen, S., 1987. Food for the spirit: vegetarianism and the world religions. New York: Bala Books, 120 pp.

Sagoff, M., 1984. Ethics and economics in environmental law. *In:* T. Regan (ed.), Earthbound: new introductory essays in environmental ethics. New York: Random House, pp. 147–178.

Schatan, J., 1987. World debt: who is going to pay? London: Zed Books, 144 pp.

Shihata, I.F.I., 1988. The World Bank and human rights: an analysis of the legal issues and the record of achievements. Denver: Denver J., 53 pp.

Singer, P., 1975. Animal liberation: a new ethics for our treatment of animals. New York: Avon Books, 297 pp.

Singer, P. (ed.), 1985. In defence of animals. Oxford: Blackwell, 224 pp.

Sivard, R., 1987. World military and social expenditures: 1986. Washington, DC: World Priorities, 52 pp.

Soderbaum, P., 1986. Economics, ethics and environmental problems. J. Interdisc. Econ. 1:139–153

Taylor, P.W., 1986. Respect for nature; a theory of environmental ethics. Princeton, NJ: Princeton University Press, 339 pp.

Uhl, C., 1986. Our steak in the jungle. BioScience 36:642.

Uhl, C., and G. Parker, 1986. Is a quarter-pound hamburger worth a half-ton of rainforest? InterCiencia 11(5):210.

U.S. Congress, 1987. A bill to protect the world's remaining tropical forests. Washington, DC: HR3010

Warford, J.J., 1986. Natural resource management and economic development. Washington, DC: World Bank, 20 pp.

Wilson, D., 1984. Pressure: the A to Z of campaigning. London: Heinemann Ed. Books, 123 pp.

World Bank, 1982. Tribal peoples and economic development: human ecologic considerations. No. 28. Washington, DC: World Bank, 288 pp.

World Resources Institute, 1985. Tropical forest action plan: a call to action. Washington, DC: World Resources Institute (with FAO, World Bank, and UNDP), 2 vols.

Worldwatch Institute, 1987. On the brink of extinction: conserving the diversity of life, paper 78. Washington, DC: Worldwatch Institute, 54 pp.

Appendix A

How to Save Tropical Forests
—a Conspectus

REDUCE DEMAND—BOOST SUPPLY

Essential Preconditions

Revamp orthodox economics; innovate Third World debt solutions; promote economic incentives for environmentally prudent behavior; assist vulnerable ethnic minorities (e.g., jungle dwellers); reduce war and its environmental effects.

Ensure conservation areas, national parks, world heritage sites, biosphere reserves; zoos and arboreta; gene, sperm, and seed banks; museums, herbaria; rehabilitate degraded tracts; promote peace, environmental education, legislation; and encourage the legal profession, e.g., pro bono; tout ethics.

Improve Development Projects

Multilateral development banks and bilateral development agencies (e.g., USAID, CIDA, SIDA, DANIDA, ODA), especially projects of forestry, rubber, oil palm, cocoa, alternatives to cocoa; promote scientific tourism, debt-for-nature swaps.

Plant rubber south of the Amazon; promote cocoa pollinators; reactivate Asian sleeping rubber; encourage tropical agro-multinationals to conserve samples of the ecosystems they use (e.g., Hershey, Dole, Firestone, Goodyear, Cadbury, Coca-Cola, Minute Maid, R.J. Reynolds' "Camel Adventure," Del Monte, Nabisco, Kentucky Fried Chicken).

Improve Land Settlement

Deflect transmigration to alang grassland, and Polonoroeste to cerrado savanna; improve land tenure in existing settlements; demote highways and cars in forest; promote carts and bicycles.

Promote cerrado, grassland, and savanna land settlements; upgrade slums; family planning; existing agriculture; fluvial transport and blimps; agrarian reform outside forest regions.

Reduce Shifting Cultivation

Except for low population-density jungle-dwelling ethnic minorities, alternatives to shifting cultivation should be encouraged.

Tropical forest action plan; improve conventional land use; promote and improve fallows and subsistence economies; promote irrigation and fertilizers on existing agricultural land.

Demote Tropical Cattle Ranching

Hamburger connection; ethics, economics, employment, sustainability (extinctions); raise health issues (e.g., cholesterol); Burger King, McDonald's.

Promote cattle on natural rangeland; promote noncattle meat (capybara, agouti); raise fish (rivers, ponds); encourage trends: red to white meat and poultry to fish to some degree of vegetarianism; cows to pigs and goats.

Improve Tropical Timber Trade

Tax tropical undressed log exports; consumer pressure against tropical lumber imports (e.g., Japan, U.S., EEC); promote sustainable forestry; reduce export incentives.

Tree plantations near markets; buffer zones around conservation areas; agroforestry; recycle paper; promote value-added to tropical wood exports.

A word of caution: violence breeds more violence and totalitarianism. Dian Fossey was macheted to death during Christmas 1984, while protecting one of our nearest nonhuman relatives, the gorilla—and there are only 252 left. On the night of July 10, 1985, a sovereign, civilized Western democracy officially sanctioned the deliberate bombing and sinking of Greenpeace's flagship *Rainbow Warrior* in Auckland harbor, killing Fernando Pereira. Reversing its denial and following two years of international arbitration, France paid Greenpeace more than 8.1 million U.S. dollars in aggravated and other damages in addition to the 7 million U.S. dollars, U.N. compensation to New Zealand, and a separate settlement with the dead victim's family. France's defense minister resigned, and the security director was fired. Nonviolence is preferable to such violence of the despoilers. The nonviolence of Mohandas K. Gandhi and

Martin Luther King are greatly preferable. The point to emphasize is that we are all needed. If your forte is golfing with chief executive officers of multinationals, then join those at the top of Appendix B. Those courageous enough to interpose their bodies between trees and chainsaws or bulldozers, or between harpoons and whales will join Greenpeace, Chipko, Sea Shepherds or Earth First. Writers, politicians, rabble-rousers, economists, media-educers, number-crunchers, fund-raisers, and bumper-sticker stickers—all are needed. Everyone has to find his own tribe. Eventually, and sooner rather than later, the nationals of tropical forest nations must manage their own resources. One highly effective action is to support conservation and environmental nongovernmental organizations (Appendix C) in such nations. This includes linking advocates of jungle dwellers with those of tropical forest, conservation, and environment.

Spectrum of Environmental Action Opportunities

(EXAMPLES ONLY)

1. *Political* (e.g., Congressmen Kasten, Obey, Porter) Hold hearings on environment and sustainability in economic development; promote legislation to promote environmental prudence (e.g., Green Party in Bundestag).

2. *Economic* (e.g., Herman Daly, Binswanger, Warford) Reduce subsidies and incentives for environmental abuse; correct pricing of externalities, etc.; promote longer term and group land tenure and intergenerational equity; promote incentives for qualitative progress rather than always quantitative growth; debt for environment equity swaps.

3. *Institutional: More Establishment*
   World Resources Institute
   Conservation Foundation
   Population Council
   International Institute for Environment and Development
   Worldwatch Institute
   IUCN CMC

4. *Institutional Grass Roots*
   Audubon Society
   National Wildlife Federation
   World Wildlife Fund
   Population-Environment Balance, Inc.

Cultural Survival
Survival International
Nature Conservancy
Sierra Club
Conservation International
Natural Resources Defense Council
Environmental Defense Fund
Defenders of Wildlife
Friends of the Earth

5. *More Activist*
Chipko Tree Huggers*
Greenpeace
Earth First
Sea Shepherds

* In 1987 Chipko received the $100,000 International Right Livelihood Award and was added to the United Nations Environmental Honor Roll, for its nonviolent resistance to environmental abuse.

Tropical Forest Action Groups

(PARTIAL LISTING ONLY)

Study, publicize, lobby, write letters, demonstrate. Active participation in some of the tropical forest conservation groups listed in Appendix B and some of the environmental groups listed in here can be exceptionally effective.

Cultural Survival, Inc., 11 Divinity Avenue, Cambridge, MA 02138, U.S.A.

Earth First, P.O. Box 5871, Tucson, AZ 85703, U.S.A.

E.F. Rainforest Action, Bay Area Earth First, P.O. Box 83, Canyon, CA 94516, U.S.A.

Environmental Liaison Center, P.O. Box 72461, Nairobi, Kenya

Friends of the Earth Campaign to Serve Tropical Rainforests, 26 Underwood Street, London N17JQ, U.K.

International Tropical Timber Organization (ITTO), Yokohama, Japan

International Union for the Conservation of Nature and Natural Resources, Avenue du Mont Blanc, CH-1196, Gland, Switzerland

Rainforest Action Network, 300 Broadway (Suite 28), San Francisco, CA 94133, U.S.A.

Rainforest Information Centre, P.O. Box 368, Lismore, NSW 2480, Australia

Sahabat Alam Malaysia, 37 Lorong Birch, 10250 Pulau Penang, Malaysia

Survival International, 2121 Decatur Place, Washington, DC 20008, U.S.A.

Tropenbos: Stimulation Program for Research in Humid Tropical Forests, Galvanistraat 9, 6716 AE Ede, Netherlands

Tropical Ecosystem Research and Rescue Alliance, P.O. Box 18391, Washington, DC 20036-8391 U.S.A.

Tropical Forest Action Network, 501 Shinwa Building, 9–17 Sakuragaoka, Shiboya-ku, Tokyo 150, Japan

Tropical Forests Working Group (c/o NRDC), 1350 New York Avenue, N.W., Washington, DC 20005, U.S.A.

Tropical Rainforest Times, FOE 26/28 Underwood St., London N1, U.K.

Tropical Forest Project, World Resources Institute, 1735 New York Avenue, N.W., Washington, DC 20006, U.S.A.

World Rainforest Report (World Rainforest Movement) (Martin Khor Kok Peng) Third World Network, 87 Cantonment Rd., 10250, Penang, Malaysia

World Wildlife Fund International, CH-1196, Gland, Switzerland

Bibliography

Biswas, A. K., and K. P. Chu. 1987. Environmental assessment for developing countries. Dublin: Tycooly Publishing, 231 pp.

Boden, R. 1982. Resource development and conservation conflicts in the Third World. Third World Planning Review 4:265–80.

Botkin, D. B., et al., eds. 1989. Changing the global environment. San Diego: Academic, 480 pp.

Carroll, C. R., and D. H. Janzen. 1973. Ecology of foraging by ants. Annual Review of Ecology Systematics 4:231–57.

Caufield, C. 1982. Tropical moist forests: the resource, the people, the threat. London: International Institute Environment and Development, 67 pp.

————. 1985. In the rainforest. New York: Knopf, 221 pp.

Chatterji, M. 1981. Energy and environment in developing countries. New York: Wiley, 357 pp.

Daly, H. E., and J. Cobb. 1989. For the common good: redirecting the economy toward community, the environment, and a sustainable future. Boston: Beacon Press, 426 pp.

Delany, M. J., and D.C.D. Happold. 1979. Ecology of African mammals. London: Longmans, 434 pp.

De Santo, R. 1978. Concepts of applied ecology. New York: Springer, 310 pp.

Deshmukh, I. 1986. Ecology and tropical biology. Palo Alto, CA: Blackwell Scientific, 387 pp.

Ehrlich, P., and A. Ehrlich. 1981. Extinction: the causes and consequences of disappearing species. New York: Random House, 331 pp.

Ehrlich, P., A. Ehrlich, and J. Holdren. 1977. Ecoscience: population, resources and environment. San Francisco: Freeman, 1052 pp.

Ewusie, J. Y. 1980. Elements of tropical ecology. London: Heinemann, 205 pp.

Furtado, J., ed. 1980. Tropical ecology and development. 2 vols. Kuala Lumpur: International Society of Tropical Ecology, 1382 pp.

Ghosh, P. K. 1984. Population, environment, resources and Third World development. Westport, CT: Greenwood Press, 634 pp.

Golley, F. B., and E. Medina. 1975. Tropical ecological systems: trends in aquatic and terrestrial research. New York: Springer, 398 pp.

Gomez Pompa, A., ed. 1975. Index of current tropical ecology research. Mexico City: Instituto Investigaciones sobre Recursos Bioticos, 255 pp.

Goodland, R., and H. Irwin. 1975. The Amazon jungle: green hell to red desert? New York: Elsevier Scientific, 150 pp.

Goodland, R., G. Ledec, and C. Watson. 1985. Environmental management in tropical agriculture. Boulder, CO: Westview Press, 236 pp.

Gradwohl, J., and R. Greenberg. 1988. Saving the tropical forests. London: Earthscan, 207 pp.

Gupta, A. 1988. Ecology and development in the Third World. London: Routledge and Kegan Paul, 80 pp.

Hartje, V. J. 1987. Environmental problems in the Third World: what can the North do? Economics 35:67–86.

Heady, H. F., and E. B. Heady. 1982. Range and wildlife management in the tropics. London: Longmans, 140 pp.

Hinckley, A. D. 1976. Applied ecology: a non-technical approach. New York: Macmillan, 342 pp.

Janzen, D. H. 1972a. Uncertain future of the tropics. Natural History 81(9):3–26.

———. 1972b. Whither tropical ecology? In Challenging biological problems, ed. J. A. Behnke, 281–296. New York: Oxford, 502 pp.

———. 1973. Tropical agroecosystems. Science 182(4118):1212–19.

———. 1974. Tropical blackwater rivers: animals and mast fruiting by the Dipterocarpaceae. Biotropica 6(2):69–103.

———. 1975. Ecology of plants in the tropics. London: Arnold, 66 pp.

———. 1976. Additional land, at what price: responsible use of the tropics in a food-population confrontation. American Phytopathology Society 3:35–39.

———. 1977. The impact of tropical studies on ecology. Philadelphia. Academy of Natural Sciences 12:159–87.

———. 1978a. The defenses of legumes against herbivores. In Advances in legume systematics, ed. R. M. Polhill and P. Raven, 951–78. Kew: Royal Botanic Gardens, 622 pp.

———. 1978b. Seeding patterns of tropical trees. In Tropical trees as living systems, ed. P. B. Tomlinson, 83–128. New York: Cambridge, 675 pp.

———. 1983a. No park is an island: increase in interference from outside as park size decreases. Oikos 41(3):402–14.

————. 1983b. Food webs: who eats what, why, how, and with what effects in a tropical forest. In Ecosystems of the world, ed. F. Golley, 167–82. New York: Elsevier, 381 pp.

————. 1983c. Dispersal of seeds by vertebrate guts. In Coevolution, ed. D. J. Futuyma, 2:32–62. Sunderland, MA: Sinauer, 555 pp.

————, ed. 1983d. Costa Rican natural history. Chicago: University of Chicago Press, 816 pp.

————. 1986a. The future of tropical ecology. Annual Review of Ecology and Systematics 17:305–24.

————. 1986b. The eternal external threat. In Conservation Biology, ed. M. Soulé, 286–303. Sunderland MA: Sinauer, 584 pp.

————. 1988. There are differences between tropical and extra-tropical parks. Oikos 51(2):121–29.

Jordan, C. F., ed. 1981. Tropical ecology. Stroudsburg, PA: Hutchinson Ross, 356 pp.

Kothari, D. V., and D. C. Ojha. 1981. Tropical ecology: an annotated bibliography. Jodhpur: Latesh Prakashan, 134 pp.

Lal, R. 1987. Tropical ecology and edaphology. New York: Wiley, 732 pp.

Ledec, G., and R. Goodland. 1988. Wildlands: their protection and management in economic development. Washington, D.C.: World Bank, 278 pp.

Leigh, E. G., ed. 1983. The ecology of a tropical forest. Washington, D.C.: Smithsonian, 468 pp.

Leonard, H. J. 1985. Divesting nature's capital: the political economy of environmental abuse in the Third World. New York: Holmes and Meier, 299 pp.

Longman, K. A., and J. Jenik. 1987. Tropical forests and its environment. New York: Wiley, 347 pp.

MacKinnon, J. et al., 1986. Managing protected areas in the tropics. Gland, Switzerland: International Union for the Conservation of Nature and Natural Resources, 295 pp.

Myers, N. 1984. The primary source: tropical forests and our future. New York: Norton, 399 pp.

NRC. 1980a. Research priorities in tropical biology. Washington, D.C.: National Research Council, 116 pp.

————. 1980b. Conversion of tropical moist forests. Washington, D.C.: National Research Council, 222 pp.

————. 1982. Ecological aspects of development in the humid tropics. Washington, D.C.: National Research Council, 297 pp.

OECD. 1989. Renewable natural resources: economic incentives for improved management. Paris: Organization for Economic Cooperation and Development, 158 pp.

Office of Technology Assessment. 1984. Technologies to sustain tropical forest. Washington, D.C.: OTA, 344 pp.

Owen, D. F. 1976. Animal ecology in tropical Africa. London: Longmans, 132 pp.

Pomeroy, D. E. 1986. Tropical ecology. New York: Longman, 233 pp.

Qasim, S. Z. 1979. Primary production in some tropical environments. In Marine Production Systems, ed. M. J. Dunbar, 31–70. New York: Cambridge University Press.

Richards, P. M. 1952. The tropical rain forest: an ecology study. Cambridge: Cambridge University Press, 587 pp.

Sutton, S. L., T. C. Whitmore, and A. C. Chadwick. 1983. Tropical rainforests: ecology and management. Oxford: Blackwell, 445 pp.

UNEP. 1982. The world environment: 1972–1982. Dublin: Tycooly Publishing, for United Nations Environment Program, 637 pp.

Viera, A da S. 1985. Environmental information in developing nations. Westport, CT: Greenwood Press, 174 pp.

Walker, B. H. 1980. Applied ecology: towards a positive approach. Johannesburg: University of Witwatersrand, Centre for Resource Ecology, 47 pp.

Whitmore, T. C. 1984. Tropical rain forests of the Far East. Oxford: Oxford University Press, 272 pp.

Contributors

ROBERT BUSCHBACHER is an associate of the Conservation Foundation (CF) and World Wildlife Fund (WWF), where he directs the Tropical Forestry Program. The CF/WWF Tropical Forestry Program conducts research and implements field projects on alternatives for productive and sustainable management of tropical forests in an effort to combine and conservation and economic development. Prior to coming to CF/WWF, Dr. Buschbacher worked at the Brazilian Humid Tropical Research Center on a Fulbright Fellowship. His research focused on the nutrient cycling, productivity, and soil fertility dynamics when Amazon rain forest is converted to cattle pasture. He has also investigated the ability of Amazon rain forest to regenerate following abandonment of pastures. He received his Ph.D. in ecology from the University of Georgia.

DR. JOSE IRENEU DOS REMEDIOS FURTADO chaired the department of Zoology at the University of Malaya since 1975 and was science adviser to the Commonwealth Secretariat from 1982 to 1987. He now directs the Centre for Integrated Development in London.

STEPHEN R. GLIESSMAN is an associate professor of environmental studies at the University of California, Santa Cruz, and director of its agroecology program. His research in agroecology and sustainable agriculture emphasizes the ecology of multiple cropping systems, allelopathy, low-input agriculture and traditional agriculture in developing countries. He has projects in Mexico and Costa Rica, and recently helped the Organization of Tropical Studies in Costa Rica establish a graduate field course in tropical agroecology. Before joining the University of California in 1981, Dr. Gliessman lived in Costa Rica and Mexico for nine years, managing farm and nursery operations for several years before joining the faculty of the Colegio Superior de Agricultura Tropical in Cardenas, Mexico, where he developed an agroecosystems research program and a master's degree curriculum in agroecology. Dr. Gliessman received his Ph.D. in plant ecology from the University of California at Santa Barbara.

CHARLES R. GOLDMAN is professor of limnology and director of the Tahoe Research Group at the Division of Environmental Studies, University of California at Davis. He has a background of 31 years of research and teaching in limnology, ecology, zoology, and aquatic biology with positions in Illinois, Alaska, and California, as well as a Fulbright professorship to Yugoslavia in 1985. He has also served on numerous federal and state committees relating to the effects of development activities on aquatic ecosystems and water quality and has worked on environmental assessments of projects in Argentina, Ecuador, and Honduras. His research since 1958 (supported by the National Science Foundation and the Environmental Protection Agency) has focused on limnological studies of lakes and reservoirs in northern California with an emphasis on eutrophication, zooplankton population dynamics, primary productivity, and limiting nutrients. Dr. Goldman received his Ph.D. in limnology and fisheries from the University of Michigan.

ROBERT GOODLAND is a tropical ecologist and chief of the newly created Environmental Division in the Latin America Office of the World Bank in Washington, D.C. He started environmental assessments of tropical development projects for the World Bank in 1972, while creating a Department of Ecology in the then new University of Brasilia. He has held a position as chairman of environmental assessment at Cary Arboretum in Millbrook, New York, as well as professorships at the University of Brasilia, INPA Manaus; the Organization of Tropical Studies, Costa Rica; and McGill University, Montreal. He has published 13 books on various topics, including environmental aspects of the Trans-Amazon Highway, the cerrado ecosystem of Brazil, tropical hydroprojects and agriculture, tribal peoples, wildland management, and cultural property. Dr. Goodland received his Ph.D. in tropical ecology from McGill University, Montreal.

AGNES KISS is an ecologist in the Environmental Division of the World Bank's Africa Regional Office, where she helps to ensure that environmental protection is an integral part of the World Bank's economic development program in that region. She also advises on pesticide and pest-management issues for all World Bank regions. This represents a continuation of her work during the preceding two years with the World Bank's Office of Environmental and Scientific Affairs, where she was responsible for monitoring the implementation of guidelines on pest-management and pesticide selection and use. Previously, she contributed to the development of those guidelines during a year as an American Association for the Advancement of Sciences, Engineering and Diplomacy Fellow in the AID Bureau for Science and Technology. Dr. Kiss received her Ph.D. in insect ecology from the University of Michigan, followed by a year of postdoctoral research at the U.S. Department of Agriculture Agricultural Research Service Laboratory in Albany, California.

FLORENCIA MONTAGNINI is academic coordinator for the Organization of Tropical Studies (OTS) at its headquarters in San José, Costa Rica, where she organizes and teaches courses in tropical ecology, agroecology, and forestry; oversees the Environmental Education Program; and works on tropical forestry research projects. She joined OTS in 1985 to work on a joint project with the Tropical Agriculture Research and Training Center (CATIE) and write a book (published in 1987) on agroforestry systems. Dr. Montagnini began her training in her home country of Argentina, graduating as an agronomist/agricultural scientist from the National University of Rosario. She received an M.S. in ecology from the Venezuelan Institute for Scientific Research, studying the influence of insects on productivity of cassava plantations in slash-and-burn agricultural sites in the Amazon. This research was part of a National Science Foundation project to study the structure and function of Amazonian rain forests and changes following perturbation and conversion to agriculture and pasture. She went on to earn a Ph.D. in ecology from the University of Georgia while working on another National Science Foundation project, mechanisms of nitrogen loss from mature and disturbed forest ecosystems, at the Coweeta Hydrologic Laboratory.

ASANAYAGUM RUDRAN is a research zoologist attached to the Department of Zoological Research, National Zoological Park, Washington, D.C. His primary area of research is the ecology and behavior of primates, and he has conducted long-term field studies on monkeys in Sri Lanka, Uganda, and Venezuela. He has also worked on elephant conservation in Sri Lanka and has advised the government on their management in areas near agricultural development projects. In 1981, Dr. Rudran initiated a program sponsored by the National Zoo to train nationals of tropical countries in wildlife conservation and management. This program provided training through short-term courses conducted overseas as well as at the National Zoo's Conservation and Research Center in Front Royal, Virginia. Thus far more than 200 wildlife biologists from 25 tropical countries have participated in the training program's seven stateside courses and 13 overseas courses. Some participants have been trained as instructors who could institutionalize the program in their home countries, while others have been assisted in obtaining higher degrees in zoology or wildlife management. Born in Sri Lanka, Dr. Rudran received his early training as a zoologist in his home country and his Ph.D. from the University of Maryland.

KATHRYN A. SATERSON is an environment and natural resources officer with the Asia/Near East Bureau of the U.S. Agency for International Development (AID). She manages the bureau's biological diversity program, assists overseas missions with environmental assessments, and works on projects related to forest, watershed, and coastal management. As an American Association for the Advancement of Science diplomacy fellow with AID's Office of Forestry, Environment and Natural Resources, Dr. Saterson provided technical assistance to AID overseas missions in Thailand, Nepal, the Philippines, and Bolivia. Dr. Saterson was the 1985–1986 AIBS Congressional Science Fellow and spent the year working for the Public Lands Subcommittee of the House Interior Committee. She also spent three years working on environmental impact assessments and natural-resources management projects for Arthur D. Little, Inc. She received a Ph.D. in biology from the University of North Carolina at Chapel Hill.

DR. MEWA SINGH, professor and head of the Zoology Department at the University of Mysore, has 10 years of experience in the conservation of Indian primates.

PETER D. VAUX is currently at the University of Nevada, Las Vegas, where he is involved in a project to fertilize 10,000 hectares of Lake Mead as the first step in a three-year experiment to enhance sports fishery. He spent three years as field director for an environmental assessment study for a hydroelectric project in Honduras, and recently returned to that country to advise the Honduran National Power Company on the development of a post-impoundment environmental program for the El Cajon reservoir. Dr. Vaux has also worked on the ecology of pelagic fishes of Lake Titicaca in Peru and Bolivia, on nutrient limitation in periphyton populations of lowland rain-forest streams in Costa Rica, and on freshwater fisheries development and subsequently on pre-feasibility environment studies for three hydroelectric projects in the Niger River basin. Dr. Vaux received his Ph.D. from the University of California, Davis.

MARYLA WEBB is a consultant. She assisted Robert Goodland with wildlands and cultural property in economic development projects. Previous experience has been as a field researcher at the Darling Center Marine Laboratory, University of Maine, and Conservation Research Center, Smithsonian Institution, and research director of Carrying Capacity, Inc., Washington, D.C., where she co-authored a monograph on trends in U.S. energy and renewable-resources demand versus supply. She received her M.F.S. in wildlife and social ecology from the Yale School of Forestry and Environmental Studies.

DR. CHRIS WEMMER is the assistant director of the U.S. National Zoological Park's Conservation and Research Center in Front Royal, Virginia. For the last 15 years he has promoted conservation research, primarily of tropical mammals and their habitats, in Latin America and Asia.

INDEX

207

Also Available from Island Press

Americans Outdoors: The Report of the President's Commission
The Legacy, The Challenge
Foreword by William K. Reilly
1987, 426 pp., appendixes, case studies, charts
Paper: $24.95 ISBN: 0-933280-36-X

The Challenge of Global Warming
Edited by Dean Edwin Abrahamson
Introduction by Senator Timothy E. Wirth
In cooperation with the Natural Resources Defense Council
1989, 350 pp., tables, graphs, index, bibliography
Cloth: $34.95 ISBN: 0-933280-87-4
Paper: $19.95 ISBN: 0-933280-86-6

Creating Successful Communities: A Guidebook to Growth
Management Strategies
By Michael A. Mantell, Stephen F. Harper, Luther Propst
In cooperation with The Conservation Foundation
1989, 350 pp., appendixes, index
Paper: $24.95 ISBN: 1-55963-014-0

Crossroads: Environmental Priorities for the Future
Edited by Peter Borrelli
1988, 352 pp., index
Cloth: $29.95 ISBN: 0-933280-68-8
Paper: $17.95 ISBN: 0-933280-67-X

Natural Resources for the 21st Century
Edited by R. Neil Sampson and Dwight Hair
In cooperation with the American Forestry Association
1989, 350 pp., index, illustrations
Cloth: $39.95 ISBN: 1-55963-003-5
Paper: $24.95 ISBN: 1-55963-002-7

Reopening the Western Frontier
From *High Country News*
1989, 350 pp., illustrations, photographs, maps, index
Cloth: $24.95 ISBN: 1-55963-011-6
Paper: $15.95 ISBN: 1-55963-010-8

Rush to Burn: Solving America's Garbage Crisis?
From *Newsday*
Winner of the Worth Bingham Award
1989, 276 pp., illustrations, photographs, graphs, index
Cloth: $29.95 ISBN: 1-55963-001-9
Paper: $14.95 ISBN: 1-55963-000-0

Saving the Tropical Forests
By Judith Gradwohl and Russell Greenberg
Preface by Michael H. Robinson
Smithsonian Institution
1988, 207 pp., index, tables, illustrations, notes, bibliography
Cloth: $24.95 ISBN: 0-933280-81-5

Shading Our Cities: Resource Guide for Urban and Community Forests
Edited by Gary Moll and Sara Ebenreck
In cooperation with the American Forestry Association
1989, 350 pp., illustrations, photographs, appendixes, index
Cloth: $34.95 ISBN: 0-933280-96-3
Paper: $19.95 ISBN: 0-933280-95-5

War on Waste: Can America Win Its Battle with Garbage?
Louis Blumberg and Robert Gottlieb
1989, 325 pp., charts, graphs, index
Cloth: $34.95 ISBN: 0-933280-92-0
Paper: $19.95 ISBN: 0-933280-91-2

Wildlife of the Florida Keys: A Natural History
By James D. Lazell, Jr.
1989, 254 pp., illustrations, photographs, line drawings, maps, index
Cloth: $31.95 ISBN: 0-933280-98-X
Paper: $19.95 ISBN: 0-933280-97-1

These titles are available from Island Press, Box 7, Covelo, CA 95428. Please enclose $2.00 shipping and handling for the first book and $1.00 for each additional book. California and Washington, D.C., residents add 6% sales tax. A catalog of current and forthcoming titles is available free of charge.